The Ultimate NSAA Guide

2nd Edition

UniAdmissions

The Ultimate NSAA Guide

400 Practice Questions

Dr. Wiraaj Agnihotri
Linh Pham
Dr. Weichao Rachel
Dr. Rohan Agarwal

About the Authors

Linh studied graduated from **Natural Sciences at St John's College, Cambridge** in 2015. A keen Physicist, Linh scored in the top 16 students nationwide in the British Physics Olympiad 2012. As a tutor, Linh specialises in helping students with science admissions tests, especially the NSAA.

After winning innovation awards from Facebook, Uber and Google, Linh is now CEO of the venture-funded LOGIVAN Technologies Pte. - a Singapore based company tackling freight transportation problems in Vietnam. Away from work, Linh relaxes by practicing yoga, swimming and badminton.

Weichao Rachel is a research fellow at University of Leeds, Faculty of Medicine and Health. She obtained her Ph.D. in Biophysics at Cambridge University after completing a B.Sc. in Physics at Stanford University where she was awarded the President's Award for Academic Excellence.

Rachel is passionate about tutoring students and has taught Physics and Engineering summer classes at Cambridge University. She has worked with UniAdmissions since 2017 – helping budding natural sciences applicants with their Oxbridge applications. She supervises physics undergraduates at Cambridge and in her free time enjoys travelling and reading detective fictions.

Wiraaj graduated in 2012 with an honours degree in Natural Sciences from Pembroke College, Cambridge where he was selected to receive annual scholarship funding from the Cambridge Trusts. In his final year at Cambridge he specialised in the rigorous Mechanisms of Disease option during which he wrote an independent research dissertation on autoimmune thyroid disease.

Wiraaj subsequently completed graduate medical studies (MBBS) at the University of Sydney in 2016. He is currently a resident medical officer in Sydney where he plans to pursue further clinical specialisation as an internal medicine physician.

With a longstanding passion for teaching, Wiraaj has been heavily involved in tutoring students at various stages of their education. Over the years Wiraaj has successfully coached a great number of students into Oxbridge. Outside of medicine and education, Wiraaj enjoys playing jazz guitar and football.

Rohan is the **Director of Operations** at *UniAdmissions* and is responsible for its technical and commercial arms. He graduated from Gonville and Caius College, Cambridge in natural sciences and is a fully qualified doctor. Over the last five years, he has tutored hundreds of successful Oxbridge and Medical applicants. He has also authored ten books on admissions tests and interviews.

Rohan has taught physiology to undergraduates and interviewed medical school applicants for Cambridge. He has published research on bone physiology and writes education articles for the Independent and Huffington Post. In his spare time, Rohan enjoys playing the piano and table tennis.

The Basics

What is the NSAA?

The Natural Sciences Admissions Assessment (NSAA) is a two hour written exam for **prospective Cambridge natural** sciences and veterinary sciences applicants.

What does the NSAA consist of?

Section	Timing	Topics Tested	Questions	Mandatory	Calculator
ONE	80 Minutes	1A: Maths 1B: Physics 1C: Chemistry 1D: Biology 1E: Advanced Maths + Physics	18 MCQs per section	Must complete Section 1A **AND two** from 1B, 1C, 1D or 1E	Not Allowed
TWO	40 Minutes	**Advanced** Biology, Chemistry & Physics	6 Long Questions	Must Complete 2 Questions from a choice of 6	Allowed

Why is the NSAA used?

Cambridge applicants tend to be a bright bunch and therefore usually have excellent grades. For example, in 2013 over 65% of students who applied to Cambridge for Natural Sciences had UMS **greater than 90% in all** of their A level subjects. This means that competition is fierce – meaning that the universities must use the NSAA to help differentiate between applicants.

When do I sit NSAA?

The NSAA takes place in the first week of November every year, normally on a Wednesday morning.

Can I resit the NSAA?

No, you can only sit the NSAA once per admissions cycle.

Where do I sit the NSAA?

You can usually sit the NSAA at your school or college (ask your exams officer for more information). Alternatively, if your school isn't a registered test centre or you're not attending a school or college, you can sit the NSAA at an authorised test centre.

Do I have to resit the NSAA if I reapply?

Yes - you cannot use your score from any previous attempts.

How is the NSAA scored?

In section 1, each question carries one mark and there is no negative marking. In section 2, marks for each question are indicated alongside it. Unless stated otherwise, you will only score marks for correct answers if you show your working.

How is the NSAA used?

Different Cambridge colleges will place different weightings on different components so it is important you find out as much information about how your marks will be used by emailing the college admissions office.

In general, the university will interview a high proportion of realistic applicants so the NSAA score isn't vital for making the interview shortlist. However, it can play a huge role in the final decision after your interview.

General Advice

Start Early

It is much easier to prepare if you practice little and often. Start your preparation well in advance; ideally by mid September but at the latest by early October. This way you will have plenty of time to complete as many papers as you wish to feel comfortable and won't have to panic and cram just before the test, which is a much less effective and more stressful way to learn. In general, an early start will give you the opportunity to identify the complex issues and work at your own pace.

Prioritise

Some questions can be long and complex – and given the intense time pressure you need to know your limits. It is essential that you don't get stuck with very difficult questions. If a question looks particularly long or complex, mark it for review and move on. You don't want to be caught 5 questions short at the end just because you took more than 3 minutes in answering a challenging multi-step maths question.

If a question is taking too long, choose a sensible answer and move on. Remember that each question carries equal weighting and therefore, you should adjust your timing accordingly. With practice and discipline, you can get very good at this and learn to maximise your efficiency.

Positive Marking

There are no penalties for incorrect answers in the NSAA; you will gain one for each right answer and will not get one for each wrong or unanswered one. This provides you with the luxury that you can always guess should you absolutely be not able to figure out the right answer for a question or run behind time. Since each question provides you with 4 to 6 possible answers, you have a 16-25% chance of guessing correctly. Therefore, if you aren't sure (and are running short of time), then make an educated guess and move on. Before 'guessing' you should try to eliminate a couple of answers to increase your chances of getting the question correct. For example, if a question has 5 options and you manage to eliminate 2 options- your chances of getting the question increase from 20% to 33%!

Avoid losing easy marks on other questions because of poor exam technique. Similarly, if you have failed to finish the exam, take the last 10 seconds to guess the remaining questions to at least give yourself a chance of getting them right.

Practice

This is the best way of familiarising yourself with the style of questions and the timing for this section. You are unlikely to be familiar with the style of questions in both sections when you first encounter them. Therefore, you want to be comfortable at using this before you sit the test.

Practising questions will put you at ease and make you more comfortable with the exam. The more comfortable you are, the less you will panic on the test day and the more likely you are to score highly. Initially, work through the questions at your own pace, and spend time carefully reading the questions and looking at any additional data. When it becomes closer to the test, **make sure you practice the questions under exam conditions**.

Past Papers

The NSAA is a very new exam so there aren't many sample papers available. Specimen papers are freely available online at www.uniadmissions.co.uk/NSAA. Once you've worked your way through the questions in this book, you are highly advised to attempt them.

Repeat Questions

When checking through answers, pay particular attention to questions you have got wrong. Study the worked solution carefully until you feel confident that you understand the reasoning, and then repeat the question without help to check that you can do it. This is the best way to learn from your mistakes, and means you are less likely to make similar mistakes when it comes to the test. The same applies for questions which you were unsure of and made an educated guess which was correct (even if you got it right). When working through this book, **make sure you highlight any questions you are unsure of**, this means you know to spend more time looking over them once marked.

Calculators

You aren't permitted to use calculators in section 1 – thus, it is essential that you have strong numerical skills. For instance, you should be able to rapidly convert between percentages, decimals and fractions. You will seldom get questions that would require calculators but you would be expected to be able to arrive at a sensible estimate. Consider for example:

Estimate 3.962 x 2.322:

3.962 is approximately 4 and 2.323 is approximately $2.33 = \frac{7}{3}$.

Thus, $3.962 \times 2.322 \approx 4 \times \frac{7}{3} = \frac{28}{3} = 9.33$

Since you will rarely be asked to perform difficult calculations, you can use this as a signpost of if you are tackling a question correctly. For example, when solving a section 1 question, you end up having to divide 8,079 by 357- this should raise alarm bells as calculations in section 1 are rarely this difficult.

It goes without saying that you should take time to familiarise yourself with your calculator's functions including the memory functions.

> *Top tip!* Don't leave things too late – do small bits early and often rather than a mad cram in the last week of October. Some of the principles tested in NSAA require a great degree of understanding and you don't do yourself justice by trying to cram them into a few hours!

A word on timing...

"If you had all day to do your NSAA, you would get 100%. But you don't."

Whilst this isn't completely true, it illustrates a very important point. Once you've practiced and know how to answer the questions, the clock is your biggest enemy. This seemingly obvious statement has one very important consequence. **The way to improve your NSAA score is to improve your speed.** There is no magic bullet. But there are a great number of techniques that, with practice, will give you significant time gains, allowing you to answer more questions and score more marks.

Timing is tight throughout the NSAA – **mastering timing is the first key to success**. Some candidates choose to work as quickly as possible to save up time at the end to check back, but this is generally not the best way to do it. NSAA questions can have a lot of information in them – each time you start answering a question it takes time to get familiar with the instructions and information. By splitting the question into two sessions (the first run-through and the return-to-check) you double the amount of time you spend on familiarising yourself with the data, as you have to do it twice instead of only once. This costs valuable time. In addition, candidates who do check back may spend 2–3 minutes doing so and yet not make any actual changes. Whilst this can be reassuring, it is a false reassurance as it is unlikely to have a significant effect on your actual score. Therefore it is usually best to pace yourself very steadily, aiming to spend the same amount of time on each question and finish the final question in a section just as time runs out. This reduces the time spent on re-familiarising with questions and maximises the time spent on the first attempt, gaining more marks.

It is essential that you don't get stuck with the hardest questions – no doubt there will be some. In the time spent answering only one of these you may miss out on answering three easier questions. If a question is taking too long, choose a sensible answer and move on. Never see this as giving up or in any way failing, rather it is the smart way to approach a test with a tight time limit. With practice and discipline, you can get very good at this and learn to maximise your efficiency. It is not about being a hero and aiming for full marks – this is almost impossible and very much unnecessary (even Cambridge doesn't expect you to get full marks!). It is about maximising your efficiency and gaining the maximum possible number of marks within the time you have.

Top tip! Ensure that you take a watch that can show you the time in seconds into the exam. This will allow you have a much more accurate idea of the time you're spending on a question. In general, if you've spent >90 seconds on a section 1 question – move on regardless of how close you think you are to solving it.

Use the Options:

Some questions may try to overload you with information. When presented with large **tables and data**, it's **essential** you look at the answer options so you can focus your mind. This can allow you to reach the correct answer a lot more quickly. Consider the example below:

The table below shows the results of a study investigating antibiotic resistance in staphylococcus populations. A single staphylococcus bacterium is chosen at random from a similar population. Resistance to any one antibiotic is independent of resistance to others.

Calculate the probability that the bacterium selected will be resistant to all four drugs.

Antibiotic	Number of Bacteria tested	Number of Resistant Bacteria
Benzyl-penicillin	10^{11}	98
Chloramphenicol	10^9	1200
Metronidazole	10^8	256
Erythromycin	10^5	2

A   1 in 10^6
B   1 in 10^{12}
C   1 in 10^{20}
D   1 in 10^{25}
E   1 in 10^{30}
F   1 in 10^{35}

Looking at the options first makes it obvious that there is **no need to calculate exact values**- only in powers of 10. This makes your life a lot easier. If you hadn't noticed this, you might have spent **well over 90 seconds** trying to calculate the exact value when it wasn't even being asked for.

In other cases, you may actually be able to use the options to arrive at the solution quicker than if you had tried to solve the question as you normally would. Consider the example below:

A region is defined by the two inequalities: $x - y^2 > 1 \ and \ xy > 1$. Which of the following points is in the defined region?

A. (10,3)
B. (10,2)
C. (-10,3)
D. (-10,2)
E. (-10,-3)

Whilst it's possible to solve this question both algebraically or graphically by manipulating the identities, by **far the quickest way is to actually use the options**. Note that options C, D and E violate the second inequality, narrowing down to answer to either A or B. For A: $10 - 3^2 = 1$ and thus this point is **on the boundary of the** defined region and not actually in the region. Thus the answer is B (as 10-4 = 6 > 1.)

In general, it pays dividends to look at the options briefly and see if they can be help you **arrive at the question more quickly**. Get into this habit early – it may feel unnatural at first but it's guaranteed **to save you time in the long run**.

Keywords

If you're stuck on a question; pay particular attention to the options that contain key modifiers like **"always"**, **"only"**, "all" as examiners like using them to test if there are any gaps in your knowledge. E.g. the statement "arteries carry oxygenated blood" would normally be true; "All arteries carry oxygenated blood" would be false because the pulmonary artery carries deoxygenated blood.

SECTION 1

Section 1 is the most time-pressured section of the NSAA. This section tests GCSE biology, chemistry, physics and maths. You have to answer 54 questions in 80 minutes. The questions can be quite difficult and it's easy to get bogged down. However, it's possible to rapidly improve if you prepare correctly so it's well worth spending time on it.

Choosing a Section

As part of section 1, you have to pick two sections from biology, chemistry, physics or advanced maths/physics. In most cases it will be immediately obvious to you which section will suit you best. Generally, applicants for physical natural sciences will choose physics/maths whilst those for biological sciences will choose the biology and chemistry. However, like the natural sciences tripos, this is by no means a hard and fast rule – it is extremely important that you choose the section you want to do ahead of time so that you can focus your preparation accordingly.

If you're unsure, take the time to review the content of each section and try out some questions so you can get a better idea of the style and difficulty of the questions. In general, the biology and chemistry questions in the NSAA require the least amount of time per question whilst the maths and physics are more time-draining as they usually consist of multi-step calculations.

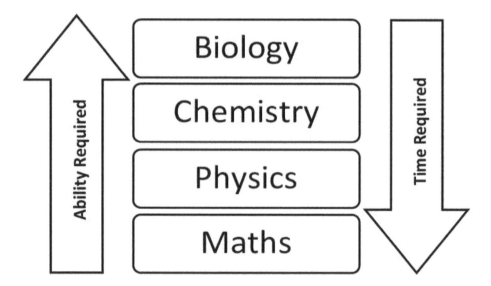

Gaps in Knowledge

The vast majority of applicants for natural sciences will be taking at least 3 science subjects. You are highly advised to go through the NSAA Specification and ensure that you have covered all examinable topics. An electronic copy of this can be obtained from **www.uniadmissions.co.uk/nsaa**.

The questions in this book will help highlight any particular areas of weakness or gaps in your knowledge that you may have. Upon discovering these, make sure you take some time to revise these topics before carrying on – there is little to be gained by attempting questions with huge gaps in your knowledge.

Maths

Being confident with maths is extremely important for the NSAA. Many students find that improving their numerical and algebraic skills usually results in big improvements in both their section 1 and 2 scores. Remember that maths in section 1 not only comes up in the maths questions but also in physics (manipulating equations and standard form) and chemistry (mass calculations). Thus, if you find yourself consistently running out of time in section 1, spending a few hours on brushing up your basic maths skills may do wonders for you.

SECTION 1A: Maths

NSAA maths questions are designed to be time draining- if you find yourself consistently not finishing, it might be worth leaving the maths (and probably physics) questions until the very end.

Good students sometimes have a habit of making easy questions difficult; remember that section 1A is pitched at GCSE level so you are not expected to know or use calculus or advanced trigonometry in it.

Formulas you **MUST** know:

2D Shapes		3D Shapes		
Area			**Surface Area**	**Volume**
Circle	πr^2	Cuboid	Sum of all 6 faces	Length x width x height
Parallelogram	Base x Vertical height	Cylinder	$2\pi r^2 + 2\pi r l$	πr^2 x 1
Trapezium	0.5 x h x (a+b)	Cone	$\pi r^2 + \pi r l$	πr^2 x (h/3)
Triangle	0.5 x base x height	Sphere	$4\pi r^2$	$(4/3)\pi r^3$

Even good students who are studying maths at A2 can struggle with certain NSAA maths topics because they're usually glossed over at school. These include:

Quadratic Formula

The solutions for a quadratic equation in the form $ax^2 + bx + c = 0$ are given by: $x = \frac{-b \pm \sqrt{b^2 - 4ac}}{2a}$

Remember that you can also use the discriminant to quickly see if a quadratic equation has any solutions:

$$If\ b^2 - 4ac < 0: No\ solutions$$
$$If\ b^2 - 4ac = 0: One\ solution$$
$$If\ b^2 - 4ac > 2: Two\ solutions$$

Completing the Square

If a quadratic equation cannot be factorised easily and is in the format $ax^2 + bx + c = 0$ then you can rearrange it into the form $a\left(x + \frac{b}{2a}\right)^2 + [c - \frac{b^2}{4a}] = 0$

This looks more complicated than it is – remember that in the NSAA, you're extremely unlikely to get quadratic equations where $a > 1$ and the equation doesn't have any easy factors. This gives you an easier equation: $\left(x + \frac{b}{2}\right)^2 + \left[c - \frac{b^2}{4}\right] = 0$ and is best understood with an example.

Consider: $x^2 + 6x + 10 = 0$

This equation cannot be factorised easily but note that: $x^2 + 6x - 10 = (x + 3)^2 - 19 = 0$

Therefore, $x = -3 \pm \sqrt{19}$. Completing the square is an important skill – make sure you're comfortable with it.

Difference between 2 Squares

If you are asked to simplify expressions and find that there are no common factors but it involves square numbers – you might be able to factorise by using the 'difference between two squares'.

For example, $x^2 - 25$ can also be expressed as $(x + 5)(x - 5)$.

Maths Questions

Question 1:

Robert has a box of building blocks. The box contains 8 yellow blocks and 12 red blocks. He picks three blocks from the box and stacks them up high. Calculate the probability that he stacks two red building blocks and one yellow building block, in **any** order.

A. $\frac{8}{20}$ B. $\frac{44}{95}$ C. $\frac{11}{18}$ D. $\frac{8}{19}$ E. $\frac{12}{20}$ F. $\frac{35}{60}$

Question 2:

Solve $\frac{3x+5}{5} + \frac{2x-2}{3} = 18$

A. 12.11 B. 13.49 C. 13.95 D. 14.2 E. 19 F. 265

Question 3:

Solve $3x^2 + 11x - 20 = 0$

A. 0.75 and $\frac{4}{3}$ C. -5 and $\frac{4}{3}$ F. -12 only

B. -0.75 and $\frac{4}{3}$ D. 5 and $\frac{4}{3}$

E. 12 only

Question 4:

Express $\frac{5}{x+2} + \frac{3}{x-4}$ as a single fraction.

A. $\frac{15x-120}{(x+2)(x-4)}$ C. $\frac{8x-14}{(x+2)(x-4)}$ F. $\frac{8x-14}{x^2-8}$

B. $\frac{8x-26}{(x+2)(x-4)}$ D. $\frac{15}{8x}$

E. 24

Question 5:

The value of p is directly proportional to the cube root of q. When p = 12, q = 27. Find the value of q when p = 24.

A. 32 B. 64 C. 124 D. 128 E. 216 F. 1728

Question 6:

Write 72^2 as a product of its prime factors.

A. $2^6 \times 3^4$ B. $2^6 \times 3^5$ C. $2^4 \times 3^4$

14

D. 2×3^3 E. $2^6 \times 3$ F. $2^3 \times 3^2$

Question 7:

Calculate: $\dfrac{2.302 \times 10^5 + 2.302 \times 10^2}{1.151 \times 10^{10}}$

A. 0.0000202   C. 0.00002002   E. 0.000002002

B. 0.00020002   D. 0.00000002   F. 0.000002002

Question 8:

Given that $y^2 + ay + b = (y + 2)^2 - 5$, find the values of **a** and **b**.

	a	b
A	-1	4
B	1	9
C	-1	-9
D	-9	1
E	4	-1
F	4	1

Question 9:

Express $\dfrac{4}{5} + \dfrac{m-2n}{m+4n}$ as a single fraction in its simplest form:

A. $\dfrac{6m+6n}{5(m+4n)}$ C. $\dfrac{20m+6n}{5(m+4n)}$ E. $\dfrac{3(3m+2n)}{5(m+4n)}$

B. $\dfrac{9m+26n}{5(m+4n)}$ D. $\dfrac{3m+9n}{5(m+4n)}$ F. $\dfrac{6m+6n}{3(m+4n)}$

Question 10:

A is inversely proportional to the square root of B. When A = 4, B = 25.

Calculate the value of A when B = 16.

A. 0.8   B. 4   C. 5   D. 6   E. 10   F. 20

Question 11:

S, T, U and V are points on the circumference of a circle, and O is the centre of the circle.

Given that angle SVU = 89°, calculate the size of the smaller angle SOU.

A. 89°   B. 91°   C. 102°   D. 178°   E. 182°   F. 212°

Question 12:

Open cylinder A has a surface area of 8π cm^2 and a volume of 2π cm^3. Open cylinder **B** is an enlargement of A and has a surface area of 32π cm^2. Calculate the volume of cylinder B.

15

A. 2π cm^3 C. 10π cm^3 E. 16π cm^3

B. 8π cm^3 D. 14π cm^3 F. 32π cm^3

Question 13:

Express $\frac{8}{x(3-x)} - \frac{6}{x}$ in its simplest form.

A. $\frac{3x-10}{x(3-x)}$ C. $\frac{6x-10}{x(3-2x)}$ E. $\frac{6x-10}{x(3-x)}$

B. $\frac{3x+10}{x(3-x)}$ D. $\frac{6x-10}{x(3+2x)}$ F. $\frac{6x+10}{x(3-x)}$

Question 14:

A bag contains 10 balls. 9 of those are white and 1 is black. What is the probability that the black ball is drawn in the tenth and final draw if the drawn balls are not replaced?

A. 0   C. $\frac{1}{100}$ D. $\frac{1}{10^{10}}$ E. $\frac{1}{362,880}$

B. $\frac{1}{10}$

Question 15:

Gambit has an ordinary deck of 52 cards. What is the probability of Gambit drawing 2 Kings (without replacement)?

A. 0   D. $\frac{4}{663}$

B. $\frac{1}{169}$ E. None of the above

C. $\frac{1}{221}$

Question 16:

I have two identical unfair dice, where the probability that the dice get a 6 is twice as high as the probability of any other outcome, which are all equally likely. What is the probability that when I roll both dice the total will be 12?

A. 0   D. $\frac{2}{7}$

B. $\frac{4}{49}$ E. None of the above

C. $\frac{1}{9}$

Question 17:

A roulette wheel consists of 36 numbered spots and 1 zero spot (i.e. 37 spots in total).

What is the probability that the ball will stop in a spot either divisible by 3 or 2?

A.   0   B. $\frac{25}{37}$ C. $\frac{25}{36}$ D. $\frac{18}{37}$ E. $\frac{24}{37}$

Question 18:

I have a fair coin that I flip 4 times. What is the probability I get 2 heads and 2 tails?

A. $\frac{1}{16}$ C. $\frac{3}{8}$

B. $\frac{3}{16}$ D. $\frac{9}{16}$

16

E. None of the above

Question 19:

Shivun rolls two fair dice. What is the probability that he gets a total of 5, 6 or 7?

A. $\frac{9}{36}$

B. $\frac{7}{12}$

C. $\frac{1}{6}$

D. $\frac{5}{12}$

E. None of the above

Question 20:

Dr Savary has a bag that contains x red balls, y blue balls and z green balls (and no others). He pulls out a ball, replaces it, and then pulls out another. What is the probability that he picks one red ball and one green ball?

A. $\frac{2(x+y)}{x+y+z}$

B. $\frac{xz}{(x+y+z)^2}$

C. $\frac{2xz}{(x+y+z)^2}$

D. $\frac{(x+z)}{(x+y+z)^2}$

E. $\frac{4xz}{(x+y+z)^4}$

F. More information necessary

Question 21:

Mr Kilbane has a bag that contains x red balls, y blue balls and z green balls (and no others). He pulls out a ball, does **NOT** replace it, and then pulls out another. What is the probability that he picks one red ball and one blue ball?

A. $\frac{2xy}{(x+y+z)^2}$

B. $\frac{2xy}{(x+y+z)(x+y+z-1)}$

C. $\frac{2xy}{(x+y+z)^2}$

D. $\frac{xy}{(x+y+z)(x+y+z-1)}$

E. $\frac{4xy}{(x+y+z-1)^2}$

F. More information needed

Question 22:

There are two tennis players. The first player wins the point with probability p, and the second player wins the point with probability 1-p. The rules of tennis say that the first player to score four points wins the game, unless the score is 4-3. At this point the first player to get two points ahead wins.

What is the probability that the first player wins in exactly 5 rounds?

A. $4p^4(1-p)$

B. $p^4(1-p)$

C. $4p(1-p)$

D. $4p(1-p)^4$

E. $4p^5(1-p)$

F. More information needed.

Question 23:

Solve the equation $\frac{4x+7}{2} + 9x + 10 = 7$

17

A. $\frac{22}{13}$ B. $-\frac{22}{13}$ C. $\frac{10}{13}$ D. $-\frac{10}{13}$ E. $\frac{13}{22}$ F. $-\frac{13}{22}$

Question 24:

The volume of a sphere is $V = \frac{4}{3}\pi r^3$, and the surface area of a sphere is $S = 4\pi r^2$. Express S in terms of V

A. $S = (4\pi)^{2/3}(3V)^{2/3}$
B. $S = (8\pi)^{1/3}(3V)^{2/3}$
C. $S = (4\pi)^{1/3}(9V)^{2/3}$
D. $S = (4\pi)^{1/3}(3V)^{2/3}$
E. $S = (16\pi)^{1/3}(9V)^{2/3}$

Question 25:

Express the volume of a cube, V, in terms of its surface area, S.

A. $V = (S/6)^{3/2}$
B. $V = S^{3/2}$
C. $V = (6/S)^{3/2}$
D. $V = (S/6)^{1/2}$
E. $V = (S/36)^{1/2}$
F. $V = (S/36)^{3/2}$

Question 26:

Solve the equations $4x + 3y = 7$ and $2x + 8y = 12$

A. $(x, y) = \left(\frac{17}{13}, \frac{10}{13}\right)$
B. $(x, y) = \left(\frac{10}{13}, \frac{17}{13}\right)$
C. $(x, y) = (1, 2)$
D. $(x, y) = (2, 1)$
E. $(x, y) = (6, 3)$
F. $(x, y) = (3, 6)$
G. No solutions possible.

Question 27:

Rearrange $\frac{(7x+10)}{(9x+5)} = 3y^2 + 2$, to make x the subject.

A. $\frac{15\,y^2}{7 - 9(3y^2+2)}$
B. $\frac{15\,y^2}{7 + 9(3y^2+2)}$
C. $-\frac{15\,y^2}{7 - 9(3y^2+2)}$
D. $-\frac{15\,y^2}{7 + 9(3y^2+2)}$
E. $-\frac{5\,y^2}{7 + 9(3y^2+2)}$
F. $\frac{5\,y^2}{7 + 9(3y^2+2)}$

Question 28:

Simplify $3x\left(\frac{3x^7}{x^{\frac{1}{3}}}\right)^3$

A. $9x^{20}$ B. $27x^{20}$ C. $87x^{20}$ D. $9x^{21}$ E. $27x^{21}$ F. $81x^{21}$

Question 29:

Simplify $2x[(2x)^7]^{\frac{1}{14}}$

A. $2x\sqrt{2\,x^4}$
B. $2x\sqrt{2x^3}$
C. $2\sqrt{2\,x^4}$
D. $2\sqrt{2x^3}$
E. $8x^3$
F. $8x$

18

Question 30:

What is the circumference of a circle with an area of 10π?

A. $2\pi\sqrt{10}$

B. $\pi\sqrt{10}$

C. 10π

D. 20π

E. $\sqrt{10}$

F. More information needed.

Question 31:

If $a.b = (ab) + (a + b),$ then calculate the value of $(3.4). 5$

A. 19   B. 54   C. 100   D. 119   E. 132

Question 32:

If $a.b = \frac{a^b}{a}$, calculate $(2.3).2$

A. $\frac{16}{3}$ B. 1   C. 2   D. 4   E. 8

Question 33:

Solve $x^2 + 3x - 5 = 0$

A. $x = -\frac{3}{2} \pm \frac{\sqrt{11}}{2}$

B. $x = \frac{3}{2} \pm \frac{\sqrt{11}}{2}$

C. $x = -\frac{3}{2} \pm \frac{\sqrt{11}}{4}$

D. $x = \frac{3}{2} \pm \frac{\sqrt{11}}{4}$

E. $x = \frac{3}{2} \pm \frac{\sqrt{29}}{2}$

F. $x = -\frac{3}{2} \pm \frac{\sqrt{29}}{2}$

Question 34:

How many times do the curves $y = x^3$ and $y = x^2 + 4x + 14$ intersect?

A. 0   B. 1   C. 2   D. 3   E. 4

Question 35:

Which of the following graphs **do not** intersect?

1. $y = x$ 2. $y = x^2$ 3. $y = 1-x^2$ 4. $y = 2$

A. 1 and 2

B. 2 and 3

C. 3 and 4

D. 1 and 3

E. 1 and 4

F. 2 and 4

Question 36:

Calculate the product of 897,653 and 0.009764.

A. 87646.8

B. 8764.68

C. 876.468

D. 87.6468

E. 8.76468

F. 0.876468

Question 37:

Solve for x: $\frac{7x+3}{10} + \frac{3x+1}{7} = 14$

A. $\frac{929}{51}$ B. $\frac{949}{47}$ C. $\frac{949}{79}$ D. $\frac{980}{79}$

Question 38:

What is the area of an equilateral triangle with side length x.

A. $\frac{x^2\sqrt{3}}{4}$ B. $\frac{x\sqrt{3}}{4}$ C. $\frac{x^2}{2}$ D. $\frac{x}{2}$ E. x^2

F. x

Question 39:

Simplify $3 - \frac{7x(25x^2-1)}{49x^2(5x+1)}$

A. $3 - \frac{5x-1}{7x}$ C. $3 + \frac{5x-1}{7x}$ E. $3 - \frac{5x^2}{49}$

B. $3 - \frac{5x+1}{7x}$ D. $3 + \frac{5x+1}{7x}$ F. $3 + \frac{5x^2}{49}$

Question 40:

Solve the equation $x^2 - 10x - 100 = 0$

A. $-5 \pm 5\sqrt{5}$ C. $5 \pm 5\sqrt{5}$ E. $5 \pm 5\sqrt{125}$

B. $-5 \pm \sqrt{5}$ D. $5 \pm \sqrt{5}$ F. $-5 \pm \sqrt{125}$

Question 41:

Rearrange $x^2 - 4x + 7 = y^3 + 2$ to make x the subject.

A. $x = 2 \pm \sqrt{y^3 + 1}$

B. $x = 2 \pm \sqrt{y^3 - 1}$

C. $x = -2 \pm \sqrt{y^3 - 1}$

D. $x = -2 \pm \sqrt{y^3 + 1}$

E. x cannot be made the subject for this equation.

Question 42:

Rearrange $3x + 2 = \sqrt{7x^2 + 2x + y}$ to make y the subject.

A. $y = 4x^2 + 8x + 2$ C. $y = 2x^2 + 10x + 2$ E. $y = x^2 + 10x + 2$

B. $y = 4x^2 + 8x + 4$ D. $y = 2x^2 + 10x + 4$ F. $y = x^2 + 10x + 4$

21

Question 43:

Rearrange $y^4 - 4y^3 + 6y^2 - 4y + 2 = x^5 + 7$ to make y the subject.

A. $y = 1 + (x^5 + 7)^{1/4}$

B. $y = -1 + (x^5 + 7)^{1/4}$

C. $y = 1 + (x^5 + 6)^{1/4}$

D. $y = -1 + (x^5 + 6)^{1/4}$

Question 44:

The aspect ratio of my television screen is 4:3 and the diagonal is 50 inches. What is the area of my television screen?

A. 1,200 inches^2

B. 1,000 inches^2

C. 120 inches^2

D. 100 inches^2

E. More information needed.

Question 45:

Rearrange the equation $\sqrt{1 + 3x^{-2}} = y^5 + 1$ to make x the subject.

A. $x = \frac{(y^{10} + 2y^5)}{3}$

B. $x = \frac{3}{(y^{10} + 2y^5)}$

C. $x = \sqrt{\frac{3}{y^{10} + 2y^5}}$

D. $x = \sqrt{\frac{y^{10} + 2y^5}{3}}$

E. $x = \sqrt{\frac{y^{10} + 2y^5 + 2}{3}}$

F. $x = \sqrt{\frac{3}{y^{10} + 2y^5 + 2}}$

Question 46:

Solve $3x - 5y = 10$ and $2x + 2y = 13$.

A. $(x, y) = (\frac{19}{16}, \frac{85}{16})$

B. $(x, y) = (\frac{85}{16}, -\frac{19}{16})$

C. $(x, y) = (\frac{85}{16}, \frac{19}{16})$

D. $(x, y) = (-\frac{85}{16}, -\frac{19}{16})$

E. No solutions possible.

Question 47:

The two inequalities $x + y \leq 3$ and $x^3 - y^2 < 3$ define a region on a plane. Which of the following points is inside the region?

A. $(2, 1)$

B. $(2.5, 1)$

C. $(1, 2)$

D. $(3, 5)$

E. $(1, 2.5)$

F. None of the above.

Question 48:

How many times do $y = x + 4$ and $y = 4x^2 + 5x + 5$ intersect?

A. 0   B. 1   C. 2   D. 3   E. 4

Question 49:

How many times do $y = x^3$ and $y = x$ intersect?

A. 0   B. 1   C. 2   D. 3   E. 4

Question 50:

A cube has unit length sides. What is the length of a line joining a vertex to the midpoint of the opposite side?

A. $\sqrt{2}$

B. $\sqrt{\frac{3}{2}}$

C. $\sqrt{3}$

D. $\sqrt{5}$

E. $\frac{\sqrt{5}}{2}$

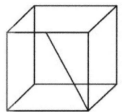

Question 51:

Solve for x, y, and z.

1. $x + y - z = -1$
2. $2x - 2y + 3z = 8$
3. $2x - y + 2z = 9$

	x	y	z
A	2	-15	-14
B	15	2	14
C	14	15	-2
D	-2	15	14
E	2	-15	14
F	No solutions possible		

Question 52:

Fully factorise: $3a^3 - 30a^2 + 75a$

A. $3a(a - 3)^3$

B. $a(3a - 5)^2$

C. $3a(a^2 - 10a + 25)$

D. $3a(a - 5)^2$

E. $3a(a + 5)^2$

23

Question 53:

Solve for x and y:

$$4x + 3y = 48$$
$$3x + 2y = 34$$

	x	y
A	8	6
B	6	8
C	3	4
D	4	3
E	30	12
F	12	30
G	No solutions possible	

Question 54:

Evaluate: $\dfrac{-\left(5^2 - 4 \times 7\right)^2}{-6^2 + 2 \times 7}$

A. $-\dfrac{3}{50}$ B. $\dfrac{11}{22}$ C. $-\dfrac{3}{22}$ D. $\dfrac{9}{50}$ E. $\dfrac{9}{22}$ F. 0

Question 55:

All license plates are 6 characters long. The first 3 characters consist of letters and the next 3 characters of numbers. How many unique license plates are possible?

A. 676,000   C. 67,600,000   E. 17,576,000

B. 6,760,000   D. 1,757,600   F. 175,760,000

Question 56:

How many solutions are there for: $2(2(x^2 - 3x)) = -9$

A. 0   D. 3

B. 1   E. Infinite solutions.

C. 2

Question 57:

Evaluate: $\left(x^{\frac{1}{2}} y^{-3}\right)^{\frac{1}{2}}$

A. $\dfrac{x^{\frac{1}{2}}}{y}$ C. $\dfrac{x^{\frac{1}{4}}}{y^{\frac{3}{2}}}$ D. $\dfrac{y^{\frac{1}{4}}}{x^{\frac{3}{2}}}$

B. $\dfrac{x}{y^{\frac{3}{2}}}$

24

Question 58:

Bryan earned a total of £ 1,240 last week from renting out three flats. From this, he had to pay 10% of the rent from the 1-bedroom flat for repairs, 20% of the rent from the 2-bedroom flat for repairs, and 30% from the 3-bedroom flat for repairs. The 3-bedroom flat costs twice as much as the 1-bedroom flat. Given that the total repair bill was £ 276 calculate the rent for each apartment.

	1 Bedroom	2 Bedrooms	3 Bedrooms
A	280	400	560
B	140	200	280
C	420	600	840
D	250	300	500
E	500	600	1,000

Question 59:

Evaluate: $5 \left[5(6^2 - 5 \times 3) + 400^{\frac{1}{2}} \right]^{1/3} + 7$

A. 0   B. 25   C. 32   D. 49   E. 56   F. 200

Question 60:

What is the area of a regular hexagon with side length 1?

A. $3\sqrt{3}$

B. $\frac{3\sqrt{3}}{2}$

C. $\sqrt{3}$

D. $\frac{\sqrt{3}}{2}$

E. 6

F. More information needed

Question 61:

Dexter moves into a new rectangular room that is 19 metres longer than it is wide, and its total area is 780 square metres. What are the room's dimensions?

A. Width = 20 m; Length = -39 m

B. Width = 20 m; Length = 39 m

C. Width = 39 m; Length = 20 m

D. Width = -39 m; Length = 20 m

E. Width = -20 m; Length = 39 m

Question 62:

Tom uses 34 meters of fencing to enclose his rectangular lot. He measured the diagonals to 13 metres long. What is the length and width of the lot?

A. 3 m by 4 m

B. 5 m by 12 m

C. 6 m by 12 m

D. 8 m by 15 m

E. 9 m by 15 m

F. 10 m by 10 m

Question 63:

Solve $\frac{3x-5}{2} + \frac{x+5}{4} = x + 1$

A. 1

B. 1.5

C. 3

D. 3.5

E. 4.5

F. None of the above

Question 64:

Calculate: $\frac{5.226 \times 10^6 + 5.226 \times 10^5}{1.742 \times 10^{10}}$

A. 0.033

B. 0.0033

C. 0.00033

D. 0.000033

E. 0.0000033

Question 65:

Calculate the area of the triangle shown to the right:

A. $3 + \sqrt{2}$

B. $\frac{2 + 2\sqrt{2}}{2}$

C. $2 + 5\sqrt{2}$

D. $3 - \sqrt{2}$

E. 3

F. 6

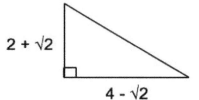

$2 + \sqrt{2}$

$4 - \sqrt{2}$

Question 66:

Rearrange $\sqrt{\frac{4}{x} + 9} = y - 2$ to make x the subject.

A. $x = \frac{11}{(y-2)^2}$

B. $x = \frac{9}{(y-2)^2}$

C. $x = \frac{4}{(y+1)(y-5)}$

D. $x = \frac{4}{(y-1)(y+5)}$

E. $x = \frac{4}{(y+1)(y+5)}$

F. $x = \frac{4}{(y-1)(y-5)}$

Question 67:

When 5 is subtracted from 5x the result is half the sum of 2 and 6x. What is the value of x?

A. 0   B. 1   C. 2   D. 3   E. 4   F. 6

Question 68:

Estimate $\frac{54.98 + 2.25^2}{\sqrt{905}}$

A. 0   B. 1   C. 2   D. 3   E. 4   F. 5

Question 69:

At a Pizza Parlour, you can order single, double or triple cheese in the crust. You also have the option to include ham, olives, pepperoni, bell pepper, meat balls, tomato slices, and pineapples. How many different types of pizza are available at the Pizza Parlour?

A. 10

B. 96

C. 192

D. 384

E. 768

F. None of the above

Question 70:

Solve the simultaneous equations $x^2 + y^2 = 1$ and $x + y = \sqrt{2}$, for x, y > 0

A. $(x,y) = (\frac{\sqrt{2}}{2}, \frac{\sqrt{2}}{2})$

B. $(x,y) = (\frac{1}{2}, \frac{\sqrt{3}}{2})$

C. $(x,y) = (\sqrt{2} - 1, 1)$

D. $(x,y) = (\sqrt{2}, \frac{1}{2})$

Question 71:

Which of the following statements is **FALSE**?

A. Congruent objects always have the same dimensions and shape.

B. Congruent objects can be mirror images of each other.

C. Congruent objects do not always have the same angles.

D. Congruent objects can be rotations of each other.

E. Two triangles are congruent if they have two sides and one angle of the same magnitude.

Question 72:

Solve the inequality $x^2 \geq 6 - x$

A. $x \leq -3$ and $x \leq 2$

B. $x \leq -3$ and $x \geq 2$

C. $x \geq -3$ and $x \leq 2$

D. $x \geq -3$ and $x \geq 2$

E. $x \geq 2$ only

F. $x \geq -3$ only

Question 73:

The hypotenuse of an isosceles right-angled triangle is x cm. What is the area of the triangle in terms of x?

A. $\frac{\sqrt{x}}{2}$

B. $\frac{x^2}{4}$

C. $\frac{x}{4}$

D. $\frac{3x^2}{4}$

E. $\frac{x^2}{10}$

Question 74:

Mr Heard derives a formula: $Q = \frac{(X+Y)^2 A}{3B}$. He doubles the values of X and Y, halves the value of A and triples the value of B. What happens to value of Q?

A. Decreases by $\frac{1}{3}$

B. Increases by $\frac{1}{3}$

C. Decreases by $\frac{2}{3}$

D. Increases by $\frac{2}{3}$

E. Increases by $\frac{4}{3}$

F. Decreases by $\frac{4}{3}$

Question 75:

Consider the graphs $y = x^2 - 2x + 3$, and $y = x^2 - 6x - 10$. Which of the following is true?

A. Both equations intersect the x-axis.

B. Neither equation intersects the x-axis.

C. The first equation does not intersect the x-axis; the second equation intersects the x-axis.

D. The first equation intersects the x-axis; the second equation does not intersect the x-axis.

SECTION 1B: Physics

Physics Questions in the NSAA are challenging as they frequently require you to make leaps in logic and calculations. Thus, before you go any further, ensure you have a firm understanding of the major principles and are confident with commonly examined topics like Newtonian mechanics, electrical circuits and radioactive decay as you may not have covered these at school depending on the specification you did.

The first step to improving in this section is to memorise by rote all the equations listed on the next page.

The majority of the physics questions involve a fair bit of maths – this means you need to be comfortable with converting between units and also powers of 10. **Most questions require two step calculations**. Consider the example:

A metal ball is released from the roof a 20 metre building. Assuming air resistance equals is negligible; calculate the velocity at which the ball hits the ground. [g = 10ms^{-2}]

A. 5 ms^{-1}
B. 10 ms^{-1}
C. 15 ms^{-1}
D. 20 ms^{-1}
E. 25 ms^{-1}

Solution: When the ball hits the ground, all of its gravitational potential energy has been converted to kinetic energy. Thus, $E_p = E_k$:

$$mg\Delta h = \frac{mv^2}{2}$$

Thus, $v = \sqrt{2gh} = \sqrt{2 \times 10 \times 20}$

$v = \sqrt{400} = 20ms^{-1}$

Here, you were required to not only recall two equations but apply and rearrange them very quickly to get the answer; all in under 60 seconds. Thus, it is easy to understand why the physics questions are generally much harder than the biology and chemistry ones.

NB: A stronger applicant would also spot that this can be solved by using a single suvat equation:

$v^2 = u^2 + 2as$

$v = \sqrt{2 \times 10 \times 20} = 20ms^{-1}$

SI Units

Remember that in order to get the correct answer you must always work in SI units i.e. do your calculations in terms of metres (not centimetres) and kilograms (not grams), etc.

Top tip! Knowing SI units is extremely useful because they allow you to **'work out' equations** if you ever forget them e.g. The units for density are kg/m^3. Since Kg is the SI unit for mass, and m^3 is represented by volume –the equation for density must be = Mass/Volume.

This can also work the other way, for example we know that the unit for Pressure is Pascal (Pa). But based on the fact that Pressure = Force/Area, a Pascal must be equivalent to N/m^2. Some physics questions will test your ability to manipulate units like this so it's important you are comfortable converting between them.

Formulas you MUST know:

Equations of Motion:

- $s = ut + 0.5at^2$
- $v = u + at$
- $a = (v-u)/t$
- $v^2 = u^2 + 2as$

Equations relating to Force:

- Force = mass x acceleration
- Force = Momentum/Time
- Pressure = Force / Area
- Moment of a Force = Force x Distance
- Work done = Force x Displacement

For objects in equilibrium:

- Sum of Clockwise moments = Sum of Anti-clockwise moments
- Sum of all resultant forces = 0

Equations relating to Energy:

- Kinetic Energy = $0.5 \, mv^2$
- Δ in Gravitational Potential Energy = $mg\Delta h$
- Energy Efficiency = (Useful energy/ Total energy) x 100%

Equations relating to Power:

- Power = Work done / time
- Power = Energy transferred / time
- Power = Force x velocity

Electrical Equations:

- $Q = It$
- $V = IR$
- $P = IV = I^2R = V^2/R$
- V = Potential difference (V, Volts)

- R = Resistance (Ohms)
- P = Power (W, Watts)
- Q = Charge (C, Coulombs)
- t = Time (s, seconds)

For Transformers: $\frac{V_p}{V_s} = \frac{n_p}{n_s}$ where:

- V: Potential difference
- n: Number of turns
- p: Primary
- s: Secondary

Other:

- Weight = mass x g
- Density = Mass / Volume
- Momentum = Mass x Velocity
- $g = 9.81 \, ms^{-2}$ (unless otherwise stated)

Factor	Text	Symbol
10^{12}	Tera	T
10^{9}	Giga	G
10^{6}	Mega	M
10^{3}	Kilo	k
10^{2}	Hecto	h
10^{-1}	Deci	d
10^{-2}	Centi	c
10^{-3}	Milli	m
10^{-6}	Micro	μ
10^{-9}	Nano	n
10^{-12}	Pico	p

Physics Questions

Question 76:

Which of the following statements are **FALSE**?

A. Electromagnetic waves cause things to heat up.
B. X-rays and gamma rays can knock electrons out of their orbits.
C. Loud sounds can make objects vibrate.
D. Wave power can be used to generate electricity.
E. Since waves carry energy away, the source of a wave loses energy.
F. The amplitude of a wave determines its mass.

Question 77:

A spacecraft is analysing a newly discovered exoplanet. A rock of unknown mass falls on the planet from a height of 30 m. Given that $g = 5.4$ ms^{-2} on the planet, calculate the speed of the rock when it hits the ground and the time it took to fall.

	Speed (ms^{-1})	Time (s)
A	18	3.3
B	18	3.1
C	12	3.3
D	10	3.7
E	9	2.3
F	1	0.3

Question 78:

A canoe floating on the sea rises and falls 7 times in 49 seconds. The waves pass it at a speed of 5 ms^{-1}. How long are the waves?

A. 12 m   B. 22 m   C. 25 m   D. 35 m   E. 57 m   F. 75 m

Question 79:

Miss Orrell lifts her 37.5 kg bike for a distance of 1.3 m in 5 s. The acceleration of free fall is 10 ms^{-2}. What is the average power that she develops?

A. 9.8 W   C. 57.9 W   E. 97.5W
B. 12.9 W   D. 79.5 W   F. 98.0 W

Question 80:

A truck accelerates at 5.6 ms^{-2} from rest for 8 seconds. Calculate the final speed and the distance travelled in 8 seconds.

	Final Speed (ms^{-1})	Distance (m)
A	40.8	119.2
B	40.8	129.6
C	42.8	179.2
D	44.1	139.2
E	44.1	179.7
F	44.2	129.2
G	44.8	179.2
H	44.8	179.7

Question 81:
Which of the following statements is true when a sky diver jumps out of a plane?

A. The sky diver leaves the plane and will accelerate until the air resistance is greater than their weight.
B. The sky diver leaves the plane and will accelerate until the air resistance is less than their weight.
C. The sky diver leaves the plane and will accelerate until the air resistance equals their weight.
D. The sky diver leaves the plane and will accelerate until the air resistance equals their weight squared.
E. The sky diver will travel at a constant velocity after leaving the plane.

Question 82:
A 100 g apple falls on Isaac's head from a height of 20 m. Calculate the apple's momentum before the point of impact. Take $g = 10$ ms^{-2}

A. 0.1 kgms^{-1}
B. 0.2 kgms^{-1}
C. 1 kgms^{-1}
D. 2 kgms^{-1}
E. 10 kgms^{-1}
F. 20 kgms^{-1}

Question 83:
Which of the following do all electromagnetic waves all have in common?

1. They can travel through a vacuum.
2. They can be reflected.
3. They are the same length.
4. They have the same amount of energy.
5. They can be polarised.

A. 1, 2 and 3 only
B. 1, 2, 3 and 4 only
C. 4 and 5 only
D. 3 and 4 only
E. 1, 2 and 5 only
F. 1 and 5 only

Question 84:
A battery with an internal resistance of 0.8 Ω and e.m.f of 36 V is used to power a drill with resistance 1 Ω. What is the current in the circuit when the drill is connected to the power supply?

A. 5 A
B. 10 A
C. 15 A
D. 20 A
E. 25 A
F. 30 A

Question 85:
Officer Bailey throws a 20 g dart at a speed of 100 ms^{-1}. It strikes the dartboard and is brought to rest in 10 milliseconds. Calculate the average force exerted on the dart by the dartboard.

A. 0.2 N
B. 2 N
C. 20 N
D. 200 N
E. 2,000 N
F. 20,000 N

Question 86:
Professor Huang lifts a 50 kg bag through a distance of 0.7 m in 3 s. What average power does she develop to 3 significant figures? Take $g = 10$ms^{-2}

A. 112 W
B. 113 W
C. 114 W
D. 115 W
E. 116 W
F. 117 W

Question 87:
An electric scooter is travelling at a speed of 30 ms^{-1} and is kept going against a 50 N frictional force by a driving force of 300 N in the direction of motion. Given that the engine runs at 200 V, calculate the current in the scooter.

A. 4.5 A
B. 45 A
C. 450 A
D. 4,500 A
E. 45,000 A
F. More information needed.

Question 88:
Which of the following statements about the physical definition of work are correct?

1. $Work\ done = \frac{Force}{distance}$
2. The unit of work is equivalent to Kgms^{-2}.
3. Work is defined as a force causing displacement of the body upon which it acts.

A. Only 1
B. Only 2
C. Only 3
D. 1 and 2
E. 2 and 3
F. 1 and 3

Question 89:
Which of the following statements about kinetic energy are correct?

1. It is defined as $E_k = \frac{mv^2}{2}$
2. The unit of kinetic energy is equivalent to Pa x m^3.
3. Kinetic energy is equal to the amount of energy needed to decelerate the body in question from its current speed.

A. Only 1
B. Only 2
C. Only 3
D. 1 and 2
E. 2 and 3
F. 1 and 3
G. 1, 2 and 3

Question 90:
In relation to radiation, which of the following statements is **FALSE**?

A. Radiation is the emission of energy in the form of waves or particles.
B. Radiation can be either ionizing or non-ionizing.
C. Gamma radiation has very high energy.
D. Alpha radiation is of higher energy than beta radiation.
E. X-rays are an example of wave radiation.

Question 91:
In relation to the physical definition of half-life, which of the following statements are correct?

1. In radioactive decay, the half-life is independent of atom type and isotope.
2. Half-life is defined as the time required for exactly half of the entities to decay.
3. Half-life applies to situations of both exponential and non-exponential decay.

A. Only 1
B. Only 2
C. Only 3
D. 1 and 2
E. 2 and 3
F. 1 and 3

Question 92:
In relation to nuclear fusion, which of the following statements is **FALSE**?

A. Nuclear fusion is initiated by the absorption of neutrons.
B. Nuclear fusion describes the fusion of hydrogen atoms to form helium atoms.
C. Nuclear fusion releases great amounts of energy.
D. Nuclear fusion requires high activation temperatures.
E. All of the statements above are false.

Question 93:
In relation to nuclear fission, which of the following statements is correct?

A. Nuclear fission is the basis of many nuclear weapons.
B. Nuclear fission is triggered by the shooting of neutrons at unstable atoms.
C. Nuclear fission can trigger chain reactions.
D. Nuclear fission commonly results in the emission of ionizing radiation.
E. All of the above.

Question 94:
Two identical resistors (R_a and R_b) are connected in a series circuit. Which of the following statements are true?

1. The current through both resistors is the same.
2. The voltage through both resistors is the same.
3. The voltage across the two resistors is given by Ohm's Law.

A. Only 1
B. Only 2
C. Only 3
D. 1 and 2

E. 2 and 3
F. 1 and 3
G. 1, 2 and 3
H. None of the statements are true.

Question 95:
The Sun is 8 light-minutes away from the Earth. Estimate the circumference of the Earth's orbit around the Sun. Assume that the Earth is in a circular orbit around the Sun. Speed of light = 3×10^8 ms^{-1}

A. 10^{24} m
B. 10^{21} m

C. 10^{18} m
D. 10^{15} m

E. 10^{12} m
F. 10^9 m

Question 96:
Which of the following statements about the physical definition of speed are true?

1. Speed is the same as velocity.
2. The internationally standardised unit for speed is ms^{-2}.
3. Velocity = distance/time.

A. Only 1
B. Only 2
C. Only 3
D. 1 and 2

E. 2 and 3
F. 1 and 3
G. 1, 2 and 3
H. None of the statements are true.

Question 97:
Which of the following statements best defines Ohm's Law?

A. The current through an insulator between two points is indirectly proportional to the potential difference across the two points.
B. The current through an insulator between two points is directly proportional to the potential difference across the two points.
C. The current through a conductor between two points is inversely proportional to the potential difference across the two points.
D. The current through a conductor between two points is proportional to the square of the potential difference across the two points.
E. The current through a conductor between two points is directly proportional to the potential difference across the two points.

Question 98:
Which of the following statements regarding Newton's Second Law are correct?

1. For objects at rest, Resultant Force must be 0 Newtons
2. Force = Mass x Acceleration
3. Force = Rate of change of Momentum

A. Only 1
B. Only 2
C. Only 3
D. 1 and 2

E. 2 and 3
F. 1 and 3
G. 1, 2 and 3

Question 99:

Which of the following equations concerning electrical circuits are correct?

1. $Charge = \frac{Voltage \times time}{Resistance}$

2. $Charge = \frac{Power \times time}{Voltage}$

3. $Charge = \frac{Current \times time}{Resistance}$

A. Only 1
B. Only 2
C. Only 3
D. 1 and 2

E. 2 and 3
F. 1 and 3
G. 1, 2 and 3
H. None of the equations are correct.

Question 100:

An elevator has a mass of 1,600 kg and is carrying passengers that have a combined mass of 200 kg. A constant frictional force of 4,000 N retards its motion upward. What force must the motor provide for the elevator to move with an upward acceleration of 1 ms^{-2}? Assume: $g = 10$ ms^{-2}

A. 1,190 N
B. 11,900 N
C. 18,000 N

D. 22,000 N
E. 23,800 N

Question 101:

A 1,000 kg car accelerates from rest at 5 ms^{-2} for 10 s. Then, a braking force is applied to bring it to rest within 20 seconds. What distance has the car travelled?

A. 125 m
B. 250 m

C. 650 m
D. 750 m

E. 1,200 m
F. More information needed

Question 102:

An electric heater is connected to 120 V mains by a copper wire that has a resistance of 8 ohms. What is the power of the heater?

A. 90 W
B. 180 W
C. 900 W
D. 1800 W

E. 9,000W
F. 18,000 W
G. More information needed

Question 103:

In a particle accelerator electrons are accelerated through a potential difference of 40 MV and emerge with an energy of 40MeV (1 MeV = 1.60 x 10^{-13} J). Each pulse contains 5,000 electrons. The current is zero between pulses. Assuming that the electrons have zero energy prior to being accelerated what is the power delivered by the electron beam?

A. 1 kW
B. 10 kW

C. 100 kW
D. 1,000 kW

E. 10,000 kW
F. More information needed

Question 104:

Which of the following statements is true?

A. When an object is in equilibrium with its surroundings, there is no energy transferred to or from the object and so its temperature remains constant.
B. When an object is in equilibrium with its surroundings, it radiates and absorbs energy at the same rate and so its temperature remains constant.
C. Radiation is faster than convection but slower than conduction.
D. Radiation is faster than conduction but slower than convection.
E. None of the above.

Question 105:
A 6kg block is pulled from rest along a horizontal frictionless surface by a constant **horizontal force of 12 N.** Calculate the speed of the block after it has moved 300 cm.

A. $2\sqrt{3}\ ms^{-1}$
B. $4\sqrt{3}\ ms^{-1}$
C. $4\sqrt{3}\ ms^{-1}$
D. $12\ ms^{-1}$
E. $\sqrt{\frac{3}{2}}\ ms^{-1}$

Question 106:
A 100 V heater heats 1.5 litres of pure water from 10°C to 50°C in 50 minutes. Given that 1 kg of pure water requires 4,000 J to raise its temperature by 1°C, calculate the resistance of the heater.

A. 12.5 ohms   C. 125 ohms   E. 500 ohms
B. 25 ohms   D. 250 ohms   F. 850 ohms

Question 107:
Which of the following statements are true?

1. Nuclear fission is the basis of nuclear energy.
2. Following fission, the resulting atoms are a different element to the original one.
3. Nuclear fission often results in the production of free neutrons and photons.

A. Only 1   E. 2 and 3
B. Only 2   F. 1 and 3
C. Only 3   G. 1, 2 and 3
D. 1 and 2   H. None of the statements are true

Question 108:
Which of the following statements are true? Assume $g = 10$ ms^{-2}.

1. Gravitational potential energy is defined as $\Delta E_p = m \times g \times \Delta h$.
2. Gravitational potential energy is a measure of the work done against gravity.
3. A reservoir situated 1 km above ground level with 10^6 litres of water has a potential energy of 1 Giga Joule.

A. Only 1   E. 2 and 3
B. Only 2   F. 1 and 3
C. Only 3   G. 1, 2 and 3
D. 1 and 2   H. None of the statements are true

Question 109:
Which of the following statements are correct in relation to Newton's 3rd law?

1. For every action there is an equal and opposite reaction.
2. According to Newton's 3rd law, there are no isolated forces.
3. Rockets cannot accelerate in deep space because there is nothing to generate an equal and opposite force.

A. Only 1   C. Only 3   E. 2 and 3
B. Only 2   D. 1 and 2   F. 1 and 3

Question 110:
Which of the following statements are correct?
1. Positively charged objects have gained electrons.
2. Electrical charge in a circuit over a period of time can be calculated if the voltage and resistance are known.
3. Objects can be charged by friction.

A. Only 1
B. Only 2
C. Only 3
D. 1 and 2

E. 2 and 3
F. 1 and 3
G. 1, 2 and 3

Question 111:
Which of the following statements is true?

A. The gravitational force between two objects is independent of their mass.
B. Each planet in the solar system exerts a gravitational force on the Earth.
C. For satellites in a geostationary orbit, acceleration due to gravity is equal and opposite to the lift from engines.
D. Two objects that are dropped from the Eiffel tower will always land on the ground at the same time if they have the same mass.
E. All of the above.
F. None of the above.

Question 112:
Which of the following best defines an electrical conductor?

A. Conductors are usually made from metals and they conduct electrical charge in multiple directions.
B. Conductors are usually made from non-metals and they conduct electrical charge in multiple directions.
C. Conductors are usually made from metals and they conduct electrical charge in one fixed direction.
D. Conductors are usually made from non-metals and they conduct electrical charge in one fixed direction.
E. Conductors allow the passage of electrical charge with zero resistance because they contain freely mobile charged particles.
F. Conductors allow the passage of electrical charge with maximal resistance because they contain charged particles that are fixed and static.

Question 113:
An 800 kg compact car delivers 20% of its power output to its wheels. If the car has a mileage of 30 miles/gallon and travels at a speed of 60 miles/hour, how much power is delivered to the wheels? 1 gallon of petrol contains 9×10^8 J.

A. 10 kW   B. 20 kW   C. 40 kW   D. 50 kW   E. 100 kW

Question 114:
Which of the following statements about beta radiation are true?

1. After a beta particle is emitted, the atomic mass number is unchanged.
2. Beta radiation can penetrate paper but not aluminium foil.
3. A beta particle is emitted from the nucleus of the atom when an electron changes into a neutron.

A. 1 only
B. 2 only

C. 1 and 3
D. 1 and 2

E. 2 and 3
F. 1, 2 and 3

Question 115:
A car with a weight of 15,000 N is travelling at a speed of 15 ms^{-1} when it crashes into a wall and is brought to rest in 10 milliseconds. Calculate the average braking force exerted on the car by the wall. Take $g = 10$ ms^{-2}

A. $1.25 \times 10^4 N$ C. $1.25 \times 10^6 N$ E. $2.25 \times 10^5 N$
B. $1.25 \times 10^5 N$ D. $2.25 \times 10^4 N$ F. $2.25 \times 10^6 N$

Question 116:
Which of the following statements are correct?

1. Electrical insulators are usually metals e.g. copper.
2. The flow of charge through electrical insulators is extremely low.
3. Electrical insulators can be charged by rubbing them together.

A. Only 1   E. 2 and 3
B. Only 2   F. 1 and 3
C. Only 3   G. 1, 2 and 3
D. 1 and 2

The following information is needed for Questions 117 and 118:

The graph below represents a car's movement. At t=0 the car's displacement was 0 m.

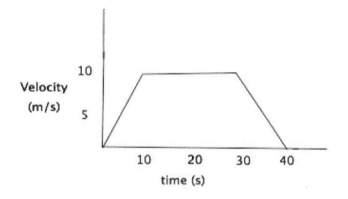

Question 117:
Which of the following statements are **NOT true**?

1. The car is reversing after t = 30.
2. The car moves with constant acceleration from t = 0 to t = 10.
3. The car moves with constant speed from t = 10 to t = 30.

A. 1 only   E. 1 and 2
B. 2 only   F. 2 and 3
C. 3 only   G. 1, 2 and 3
D. 1 and 3

Question 118:
Calculate the distance travelled by the car.

A. 200 m   C. 350 m   E. 500 m
B. 300 m   D. 400 m   F. More information needed

~ 37 ~

Question 119:

A 1,000 kg rocket is launched during a thunderstorm and reaches a constant velocity 30 seconds after launch. Suddenly, a strong gust of wind acts on it for 5 seconds with a force of 10,000 N in the direction of movement. What is the resulting change in velocity?

A. 0.5 ms^{-1} C. 50 ms^{-1} E. 5000 ms^{-1}

B. 5 ms^{-1} D. 500 ms^{-1} F. More information needed

Question 120:

A 0.5 tonne crane lifts a 0.01 tonne wardrobe by 100 cm in 5,000 milliseconds.
Calculate the average power developed by the crane. Take $g = 10 \text{ ms}^{-2}$.

A. 0.2 W   C. 5 W   E. 50 W

B. 2 W   D. 20 W   F. More information needed

Question 121:

A 20 V battery is connected to a circuit consisting of a 1 Ω and 2 Ω resistor in parallel. Calculate the overall current of the circuit.

A. 6.67 A   B. 8 A   C. 10 A   D. 12 A   E. 20 A   F. 30 A

Question 122:

Which of the following statements is correct?

A. The speed of light changes when it enters water.
B. The speed of light changes when it leaves water.
C. The direction of light changes when it enters water.
D. The direction of light changes when it leaves water.
E. All of the above.
F. None of the above.

Question 123:

In a parallel circuit, a 60 V battery is connected to two branches. Branch A contains 6 identical 5 Ω resistors and branch B contains 2 identical 10 Ω resistors.

Calculate the current in branches A and B.

	I_A (A)	I_B (A)
A	0	6
B	6	0
C	2	3
D	3	2
E	3	3
F	1	5
G	5	1

Question 124:
Calculate the voltage of an electrical circuit that has a power output of 50,000,000,000 nW and a current of 0.000000004 GA.

A. 0.0125 GV
B. 0.0125 MV
C. 0.0125 kV
D. 0.0125 V

E. 0.0125 mV
F. 0.0125 μV
G. 0.0125 nV

Question 125:
Which of the following statements about radioactive decay is correct?

A. Radioactive decay is highly predictable.
B. An unstable element will continue to decay until it reaches a stable nuclear configuration.
C. All forms of radioactive decay release gamma rays.
D. All forms of radioactive decay release X-rays.
E. An atom's nuclear charge is unchanged after it undergoes alpha decay.
F. None of the above.

Question 126:
A circuit contains three identical resistors of unknown resistance connected in series with a 15 V battery. The power output of the circuit is 60 W.
Calculate the overall resistance of the circuit when two further identical resistors are added to it.

A. 0.125 Ω
B. 1.25 Ω

C. 3.75 Ω
D. 6.25 Ω

E. 18.75 Ω
F. More information needed.

Question 127:
A 5,000 kg tractor's engine uses 1 litre of fuel to move 0.1 km. 1 ml of the fuel contains 20 kJ of energy.
Calculate the engine's efficiency. Take $g = 10$ ms^{-2}

A. 2.5 %
B. 25 %

C. 38 %
D. 50 %

E. 75 %
F. More information needed.

Question 128:
Which of the following statements are correct?

1. Electromagnetic induction occurs when a wire moves relative to a magnet.
2. Electromagnetic induction occurs when a magnetic field changes.
3. An electrical current is generated when a coil rotates in a magnetic field.

A. Only 1
B. Only 2
C. Only 3
D. 1 and 2

E. 2 and 3
F. 1 and 3
G. 1, 2 and 3

Question 129:
Which of the following statements are correct regarding parallel circuits?
1. The current flowing through a branch is dependent on the branch's resistance.
2. The total current flowing into the branches is equal to the total current flowing out of the branches.
3. An ammeter will always give the same reading regardless of its location in the circuit.

A. Only 1
B. Only 2
C. Only 3
D. 1 and 2

E. 2 and 3
F. 1 and 3
G. All of the above

Question 130:
Which of the following statements regarding series circuits are true?
1. The overall resistance of a circuit is given by the sum of all resistors in the circuit.
2. Electrical current moves from the positive terminal to the negative terminal.
3. Electrons move from the positive terminal to the negative terminal.

A. Only 1
B. Only 2

C. Only 3
D. 1 and 2

E. 2 and 3
F. 1 and 3

Question 131:
The graphs below show current vs. voltage plots for 4 different electrical components.

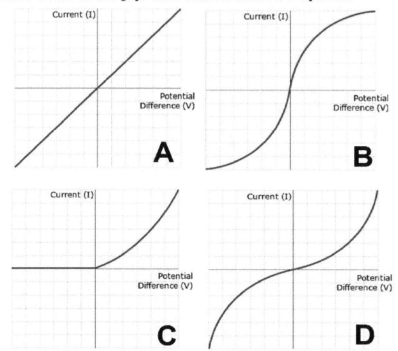

Which of the following graphs represents a resistor at constant temperature, and which a filament lamp?

	Fixed Resistor	Filament Lamp
A	A	B
B	A	C
C	A	D
D	C	A
E	C	C
F	C	D

Question 132:

Which of the following statements are true about vectors?

A. Vectors can be added or subtracted.
B. All vector quantities have a defined magnitude.
C. All vector quantities have a defined direction.
D. Displacement is an example of a vector quantity.
E. All of the above.
F. None of the above.

Question 133:
The acceleration due to gravity on the Earth is six times greater than that on the moon. Dr Tyson records the weight of a rock as 250 N on the moon.

Calculate the rock's density given that it has a volume of 250 cm³. Take $g_{Earth} = 10$ ms^{-2}
A. 0.2 kg/cm³   C. 0.6 kg/cm³   E. 0.8 kg/cm³
B. 0.5 kg/cm³   D. 0.7 kg/cm³   F. More information needed.

Question 134:
A radioactive element X_{78}^{225} undergoes alpha decay. What is the atomic mass and atomic number after 5 alpha particles have been released?

	Mass Number	Atomic Number
A	200	56
B	200	58
C	205	64
D	205	68
E	215	58
F	215	73
G	225	78
H	225	83

Question 135:
A 20 A current passes through a circuit with resistance of 10 Ω. The circuit is connected to a transformer that contains a primary coil with 5 turns and a secondary coil with 10 turns. Calculate the potential difference exiting the transformer.

A. 100 V   E. 2,000 V
B. 200 V   F. 4,000 V
C. 400 V   G. 5,000 V
D. 500 V

Question 136:
A metal sphere of unknown mass is dropped from an altitude of 1 km and reaches terminal velocity 300 m before it hits the ground. Given that resistive forces do a total of 10 kJ of work for the last 100 m before the ball hits the ground, calculate the mass of the ball. Take $g = 10$ms^{-2}.

A. 1 kg   C. 5 kg   E. 20 kg
B. 2 kg   D. 10 kg   F. More information needed.

Question 137:
Which of the following statements is true about the electromagnetic spectrum?

A. The wavelength of ultraviolet waves is shorter than that of x-rays.
B. For waves in the electromagnetic spectrum, wavelength is directly proportional to frequency.
C. Most electromagnetic waves can be stopped with a thin layer of aluminium.
D. Waves in the electromagnetic spectrum travel at the speed of sound.
E. Humans are able to visualise the majority of the electromagnetic spectrum.
F. None of the above.

Question 138:
In relation to the Doppler Effect, which of the following statements are true?

1. If an object emitting a wave moves towards the sensor, the wavelength increases and frequency decreases.
2. An object that originally emitted a wave of a wavelength of 20 mm followed by a second reading delivering a wavelength of 15 mm is moving towards the sensor.
3. The faster the object is moving away from the sensor, the greater the increase in frequency.

A. Only 1
B. Only 2
C. Only 3
D. 1 and 2

E. 1 and 3
F. 2 and 3
G. 1, 2 and 3
H. None of the above statements are true.

Question 139:
A 5 g bullet is travels at 1 km/s and hits a brick wall. It penetrates 50 cm before being brought to rest 100 ms after impact. Calculate the average braking force exerted by the wall on the bullet.

A. 50 N
B. 500 N

C. 5,000 N
D. 50,000 N

E. 500,000 N
F. More information needed.

Question 140:
Polonium (Po) is a highly radioactive element that has no known stable isotope. Po^{210} undergoes radioactive decay to Pb^{206} and Y. Calculate the number of protons in 10 moles of Y. [Avogadro's Constant = 6×10^{23}]

A. 0
B. 1.2×10^{24}

C. 1.2×10^{25}
D. 2.4×10^{24}

E. 2.4×10^{25}
F. More information needed

Question 141:
Dr Sale measures the background radiation in a nuclear wasteland to be 1,000 Bq. He then detects a spike of 16,000 Bq from a nuclear rod made up of an unknown material. 300 days later, he visits and can no longer detect a reading higher than 1,000 Bq from the rod, even though it hasn't been disturbed.
What is the longest possible half-life of the nuclear rod?

A. 25 days
B. 50 days

C. 75 days
D. 100 days

E. 150 days
F. More information needed

Question 142:

A radioactive element Y_{89}^{200} undergoes a series of beta (β^-) and gamma decays. What are the number of protons and neutrons in the element after the emission of 5 beta particles and 2 gamma waves?

	Protons	Neutrons
A	79	101
B	84	111
C	84	116
D	89	111
E	89	106
F	94	111
G	94	106
H	109	111

Question 143:

Most symphony orchestras tune to 'standard pitch' (frequency = 440 Hz). When they are tuning, sound directly from the orchestra reaches audience members that are 500 m away in 1.5 seconds.
Estimate the wavelength of 'standard pitch'.

A. 0.05 m   C. 0.75 m   E. 15 m
B. 0.5 m   D. 1.5 m   F. More information needed

Question 144:

A 1 kg cylindrical artillery shell with a radius of 50 mm is fired at a speed of 200 ms^{-1}. It strikes an armour plated wall and is brought to rest in 500 μs.

Calculate the average pressure exerted on the entire shell by the wall at the time of impact.

A. 5×10^6 Pa   C. 5×10^8 Pa   E. 5×10^{10} Pa
B. 5×10^7 Pa   D. 5×10^9 Pa   F. More information needed

Question 145:

A 1,000 W display fountain launches 120 litres of water straight up every minute. Given that the fountain is 10% efficient, calculate the maximum possible height that the stream of water could reach.
Assume that there is negligible air resistance and $g = 10$ ms^{-2}.

A. 1 m   C. 10 m   E. 50m
B. 5 m   D. 20 m   F. More information needed

Question 146:

In relation to transformers, which of the following is true?
1. Step up transformers increase the voltage leaving the transformer.
2. In step down transformers, the number of turns in the primary coil is smaller than in the secondary coil.
3. For transformers that are 100% efficient: $I_p V_p = I_s V_s$

A. Only 1   E. 1 and 3
B. Only 2   F. 2 and 3
C. Only 3   G. 1, 2 and 3
D. 1 and 2   H. None of the above.

Question 147:

The half-life of Carbon-14 is 5,730 years. A bone is found that contains 6.25% of the amount of C^{14} that would be found in a modern one. How old is the bone?

A. 11,460 years
B. 17,190 years
C. 22,920 years
D. 28,650 years
E. 34,380 years
F. 40,110 years

Question 148:

A wave has a velocity of 2,000 mm/s and a wavelength of 250 cm. What is its frequency in MHz?

A. 8×10^{-3} MHz
B. 8×10^{-4} MHz
C. 8×10^{-5} MHz
D. 8×10^{-6} MHz
E. 8×10^{-7} MHz
F. 8×10^{-8} MHz

Question 149:

A radioactive element has a half-life of 25 days. After 350 days it has a count rate of 50. What was its original count rate?

A. 102,400
B. 162,240
C. 204,800
D. 409,600
E. 819,200
F. 1,638,400
G. 3,276,800

Question 150:

Which of the following units is **NOT** equivalent to a Volt (V)?

A. $A\Omega$
B. WA^{-1}
C. $Nms^{-1}A^{-1}$
D. NmC
E. JC^{-1}
F. $JA^{-1}s^{-1}$

SECTION 1C: Chemistry

Most students don't struggle with NSAA chemistry - however, there are certain questions that even good students tend to struggle with under time pressure e.g. balancing equations and mass calculations. It is essential that you're able to do these quickly as they take up by far the most time in the chemistry questions.

Balancing Equations

For some reason, most students are rarely shown how to formally balance equations – including those studying it at A-level. Balancing equations intuitively or via trial and error will only get you so far in the NSAA as the equations you'll have to work with will be fairly complex. To avoid wasting valuable time, it is essential you learn a method that will allow you to solve these in less than 60 seconds on a consistent basis. The method shown below is the simplest way and requires you to be able to do quick mental arithmetic (which is something you should be aiming for anyway). The easiest way to do learn it is through an example:

The following equation shows the reaction between Iodic acid, hydrochloric acid and copper Iodide:

$$\text{a } HIO_3 + \text{b } CuI_2 + \text{c } HCl \rightarrow \text{d } CuCl_3 + \text{e } ICl + \text{f } H_2O$$

What values of **a**, **b**, **c**, **d**, **e** and **f** are needed in order to balance the equation?

	a	b	c	d	e	f
A	5	4	25	4	13	15
B	5	4	20	4	8	15
C	5	6	20	6	8	15
D	2	8	10	8	8	15
E	6	8	24	10	16	15
F	6	10	22	10	16	15

Step 1: Pick an element and see how many atoms there are on the left and right sides.

Step 2: Form an equation to represent this. For Cu: b = d

Step 3: See if any of the answer options don't satisfy b=d. In this case, for option **E**, b is 8 and d is 10. This allows us to eliminate option E.

Once you've eliminated as many options as possible, go back to step 1 and pick another element.
For Hydrogen (H): a + c = 2f

Then see if any of the answer options don't satisfy a + c = 2f.

➤ Option **A**: 5 + 25 is equal to 2 x 15
➤ Option **B**: 5 + 20 is not equal to 2 x 15
➤ Option **C**: 5 + 20 is not equal to 2 x 15
➤ Option **D**: 2 + 10 is not equal to 2 x 15

This allows us to eliminate option **B**, **C** & **D**. Since **E** was eliminated earlier, **A** is the only possible solution. This method works best when you get given a table above as this allows you to quickly eliminate options. However, it is still a viable method even if you don't get this information.

Chemistry Calculations

Equations you **MUST** know:

- Number of moles $n = \frac{m}{M_r}$

- Amount (mol) = Concentration (mol/dm³) x Volume (dm³)

- Gas law $pV = NRT$

- Enthalpy of formation $\Delta H_f^{\ominus} = \Delta H_{products}^{\ominus} - \Delta H_{reactants}^{\ominus}$

- Enthalpy of combustion $\Delta H_c^{\ominus} = \Delta H_{reactants}^{\ominus} - \Delta H_{products}^{\ominus}$

- pH formula $pH = -log_{10}([\text{H}^+])$

- Equilibrium $[A] + [B] = [C] + [D]$

- Equilibrium constant $10^{-pK} = \frac{[A][B]}{[C][D]}$

Essential constants

- Atomic mass constant u = 1.66×10^{-27} kg
- Atomic mass of Hydrogen = 1u
- Atomic mass of Carbon = 12u
- Avogadro's constant $R_a = 6.023 \times 10^{23}$

Avogadro's Constant:

One mole of anything contains 6×10^{23} of it e.g. 5 Moles of water contain $5 \times 6 \times 10^{23}$ number of water molecules.

Abundances:

The average atomic mass takes the abundances of all isotopes into account. Thus:

A_r = (Abundance of Isotope 1) x (Mass of Isotope 1) + (Abundance of Isotope 2) x (Mass of Isotope 2) +...

It's easier to understand this by working through examples e.g. **questions 406, 412 and 439**.

Top tip! Ensure you're able convert between **Litres, dm³, cm³ and mm³** quickly and accurately so that you don't make silly mistakes in the real exam when under time pressure.

Chemistry Questions

Question 151:
Which of the following most accurately defines an isotope?

A. An isotope is an atom of an element that has the same number of protons in the nucleus but a different number of neutrons orbiting the nucleus.
B. An isotope is an atom of an element that has the same number of neutrons in the nucleus but a different number of protons orbiting the nucleus.
C. An isotope is any atom of an element that can be split to produce nuclear energy.
D. An isotope is an atom of an element that has the same number of protons in the nucleus but a different number of neutrons in the nucleus.
E. An isotope is an atom of an element that has the same number of protons in the nucleus but a different number of electrons orbiting it.

Question 152:
Which of the following is an example of a displacement reaction?

1. $Fe + SnSO_4 \rightarrow FeSO_4 + Sn$
2. $Cl_2 + 2KBr \rightarrow Br_2 + 2KCl$
3. $H_2SO_4 + Mg \rightarrow MgSO_4 + H_2$
4. $Pb(NO_3)_2 + 2NaCl \rightarrow PbCl_2 + 2NaNO_3$

A. 1 only
B. 1 and 2 only
C. 2 and 3 only
D. 3 and 4 only
E. 1, 2 and 3 only
F. 1,2, 3 and 4 only

Question 153:
What values of **a**, **b** and **c** are needed to balance the equation below?

$$aCa(OH)_2 + bH_3PO_4 \rightarrow Ca_3(PO_4)_2 + cH_2O$$

A. $a = 3$ $b = 2$ $c = 6$
B. $a = 2$ $b = 2$ $c = 4$
C. $a = 3$ $b = 2$ $c = 1$
D. $a = 1$ $b = 2$ $c = 3$
E. $a = 4$ $b = 2$ $c = 6$
F. $a = 3$ $b = 2$ $c = 4$

Question 154:
What values of **s**, **t** and **u** are needed to balance the equation below?

$$sAgNO_3 + tK_3PO_4 \rightarrow 3Ag_3PO_4 + uKNO_3$$

A. $s = 9$ $t = 3$ $u = 9$
B. $s = 6$ $t = 3$ $u = 9$
C. $s = 9$ $t = 3$ $u = 6$
D. $s = 9$ $t = 6$ $u = 9$
E. $s = 3$ $t = 3$ $u = 9$
F. $s = 9$ $t = 3$ $u = 3$

Question 155:
Which of the following statements are true with regard to displacement?

1. A less reactive halogen can displace a more reactive halogen.
2. Chlorine cannot displace bromine or iodine from an aqueous solution of its salts.
3. Bromine can displace iodine because of the trend of reactivity.
4. Fluorine can displace chlorine as it is higher up the group.
5. Lithium can displace francium as it is higher up the group.

A. 3 only
B. 5 only
C. 1 and 2 only

D. 3 and 4 only
E. 2 , 3 and 5 only
F. 3, 4 and 5 only

Question 156:
What mass of magnesium oxide is produced when 75g of magnesium is burned in excess oxygen?
Relative Atomic Masses: Mg = 24, O = 16

A. 80g   B. 100g   C. 125g   D. 145g   E. 175g   F. 225g

Question 157:
Hydrogen can combine with hydroxide ions to produce water. Which process is involved in this?

A. Hydration
B. Oxidation

C. Reduction
D. Dehydration

E. Evaporation
F. Precipitation

Question 158:
Which of the following statements about Ammonia are correct?

1. It has a formula of NH_3.
2. Nitrogen contributes 82% to its mass.
3. It can be broken down again into nitrogen and hydrogen.
4. It is covalently bonded.
5. It is used to make fertilisers.

A. 1 and 2 only
B. 1 and 4 only
C. 1, 2 and 3 only

D. 1, 2 and 5 only
E. 3, 4 and 5 only
F. 1, 2, 3, 4 and 5

Question 159:
What colour will a universal indicator change to in a solution of milk and lipase?

A. From green to orange.
B. From red to green.
C. From purple to green.

D. From purple to orange.
E. From yellow to purple.
F. From purple to red.

Question 160:
Vitamin C [$C_6H_8O_6$] can be artificially synthesised from glucose [$C_6H_{12}O_6$]. What type of reaction is this likely to be?

A. Dehydration
B. Hydration

C. Oxidation
D. Reduction

E. Displacement
F. Evaporation

Question 161:
Which of the following statements are true?

1. Cu^{64} will undergo oxidation faster than Cu^{65}.
2. Cu^{65} will undergo reduction faster than Cu^{64}.
3. Cu^{65} and Cu^{64} have the same number of electrons.

A. 1 only   C. 3 only   E. 1 and 3 only
B. 2 only   D. 2 and 3 only   F. 1, 2 and 3

Question 162:
6g of Mg^{24} is added to a solution containing 30g of dissolved sulphuric acid (H_2SO_4). Which of the following statements are true?
Relative Atomic Masses: S = 32, Mg = 24, O = 16, H = 1

1. In this reaction, the magnesium is the limiting reagent
2. In this reaction, sulphuric acid is the limiting reagent
3. The mass of salt produced equals the original mass of sulphuric acid

A. 1 only   C. 3 only   E. 1 and 3 only
B. 2 only   D. 1 and 2 only   F. 2 and 3 only

Question 163:
In which of the following mixtures will a displacement reaction occur?

1. $Cu + 2AgNO_3$
2. $Cu + Fe(NO_3)_2$
3. $Ca + 2H_2O$
4. $Fe + Ca(OH)_2$

A. 1 only   C. 3 only   E. 1 and 2 only   G. 1, 2 and 3
B. 2 only   D. 4 only   F. 1 and 3 only   H. 1, 2, 3 and 4

Question 164:
Which of the following statements is true about the following chain of metals?

$$Na \rightarrow Ca \rightarrow Mg \rightarrow Al \rightarrow Zn$$

Moving from left to right:

1. The reactivity of the metals increases.
2. The likelihood of corrosion of the metals increases.
3. More energy is required to separate these metals from their ores.
4. The metals lose electrons more readily to form positive ions.

A. 1 and 2 only
B. 1 and 3 only
C. 2 and 3 only
D. 1 and 4 only
E. 2, 3 and 4 only
F. 1, 2, 3 and 4
G. None of the statements is correct.

Question 165:
In which of the following mixtures will a displacement reaction occur?

1. $I_2 + 2KBr$
2. $Cl_2 + 2NaBr$
3. $Br_2 + 2KI$

A. 1 only
B. 2 only
C. 3 only
D. 1 and 2 only
E. 1 and 3 only
F. 2 and 3 only
G. 1, 2 and 3

Question 166:
Which of the following statements about Al and Cu are true?

1. Al is used to build aircraft because it is lightweight and resists corrosion.
2. Cu is used to build electrical wires because it is a good insulator.
3. Both Al and Cu are good conductors of heat.
4. Al is commonly alloyed with other metals to make coins.
5. Al is resistant to corrosion because of a thin layer of aluminium hydroxide on its surface.

A. 1 and 3 only
B. 1 and 4 only
C. 1, 3 and 5 only
D. 1, 3, 4, 5 only
E. 2, 4 and 5 only
F. 2, 3, 4, 5 only

Question 167:
21g of Li^7 reacts completely with excess water. Given that the molar gas volume is 24 dm^3 under the conditions, what is the volume of hydrogen produced?

A. 12 dm^3
B. 24 dm^3
C. 36 dm^3
D. 48 dm^3
E. 72 dm^3
F. 120 dm^3

Question 168:
Which of the following statements regarding bonding are true?

1. NaCl has stronger ionic bonds than $MgCl_2$.
2. Transition metals are able to lose varying numbers of electrons to form multiple stable positive ions.
3. All covalently bonded structures have lower melting points than ionically bonded compounds.
4. All covalently bonded structures do not conduct electricity.

A. 1 only
B. 2 only
C. 3 only
D. 4 only
E. 1 and 2 only
F. 2 and 3 only
G. 3 and 4 only
H. 1, 2 and 4 only

Question 169:
Consider the following two equations:

A.	$C + O_2 \rightarrow CO_2$	$\Delta H = -394$ kJ per mole
B.	$CaCO_3 \rightarrow CaO + CO_2$	$\Delta H = + 178$ kJ per mole

Which of the following statements are true?

1.   Reaction **A** is exothermic and Reaction **B** is endothermic.
2. CO_2 has less energy than C and O_2.
3.   CaO is more stable than $CaCO_3$.

A.   1 only
B.   2 only
C.   3 only
D.   1 and 2
E.   1 and 3
F.   2 and 3
G.   1, 2 and 3

Question 170:
Which of the following are true of regarding the oxides formed by Na, Mg and Al?

1.   All of the metals and their solid oxides conduct electricity.
2.   MgO has stronger bonds than Na_2O.
3.   Metals are extracted from their molten ores by fractional distillation.

A.	1 only	C.	3 only	E.	2 and 3 only
B.	2 only	D.	1 and 2 only	F.	1, 2 and 3

Question 171:
Which of the following pairs have the same electronic configuration?

1. Li^+ and Na^+
2. Mg^{2+} and Ne
3. Na^{2+} and Ne
4. O^{2-} and a Carbon atom

A.	1 only	C.	1 and 3 only	E.	2 and 4 only
B.	1 and 2 only	D.	2 and 3 only	F.	1, 2, 3 and 4

Question 172:
In relation to reactivity of elements in group 1 and 2, which of the following statements **is correct?**

1.   Reactivity decreases as you go down group 1.
2.   Reactivity increases as you go down group 2.
3.   Group 1 metals are generally less reactive than group 2 metals.

A.	Only 1	C.	Only 3	E.	2 and 3
B.	Only 2	D.	1 and 2	F.	1 and 3

Question 173:
What role do catalysts fulfil in an endothermic reaction?

A. They increase the temperature, causing the reaction to occur at a faster rate.
B. They decrease the temperature, causing the reaction to occur at a faster rate.
C. They reduce the energy of the reactants in order to trigger the reaction.
D. They reduce the activation energy of the reaction.
E. They increase the activation energy of the reaction.

Question 174:
Tritium H^3 is an isotope of Hydrogen. Why is tritium commonly referred to as 'heavy hydrogen'.

A. Because H^3 contains 3 protons making it heavier than H^1 that contains 1 proton.
B. Because H^3 contains 3 neutrons making it heavier than H^1 that contains 1 neutron.
C. Because H^3 contains 1 neutron and 2 protons making it heavier than H^1 that contains 1 neutron and 1 proton.
D. Because H^3 contains 1 proton and 2 neutrons making it heavier than H^1 that contains 1 proton.
E. Because H^3 contains 3 electrons making it heavier than H^1 that contains 1 electron.

Question 175:
In relation to redox reactions, which of the following statements are correct?

1. Oxidation describes the loss of electrons.
2. Reduction increases the electron density of an ion, atom or molecule.
3. Halogens are powerful reducing agents.

A. Only 1   C. Only 3   E. 2 and 3
B. Only 2   D. 1 and 2   F. 1 and 3

Question 176:
Which of the following statements is correct?

A. At higher temperatures, gas molecules move at angles that cause them to collide with each other more frequently.
B. Gas molecules have lower energy after colliding with each other.
C. At higher temperatures, gas molecules attract each other resulting in more collisions.
D. The average kinetic energy of gas molecules is the same for all gases at the same temperature.
E. The momentum of gas molecules decreases as pressure increases.

Question 177:
Which of the following are exothermic reactions?

1. Burning Magnesium in pure oxygen
2. The combustion of hydrogen
3. Aerobic respiration
4. Evaporation of water in the oceans
5. Reaction between a strong acid and a strong base

A. 1, 2 and 4   C. 1, 3 and 5   E. 1, 2, 3 and 5
B. 1, 2 and 5   D. 2, 3 and 4   F. 1, 2, 3, 4 and 5

Question 178:

Ethene reacts with oxygen to produce water and carbon dioxide. Which elements are oxidised/reduced?

A. Carbon is reduced and oxygen is oxidised.
B. Hydrogen is reduced and oxygen is oxidised.
C. Carbon is oxidised and hydrogen is reduced.
D. Hydrogen is oxidised and carbon is reduced.
E. Carbon is oxidised and oxygen is reduced.
F. None of the above.

Question 179:

In the reaction between Zinc and Copper (II) sulphate which elements act as oxidising + reducing agents?

A. Zinc is the reducing agent while sulfur is the oxidizing agent.
B. Zinc is the reducing agent while copper in $CuSO_4$ is the oxidizing agent.
C. Copper is the reducing agent while zinc is the oxidizing agent.
D. Oxygen is the reducing agent while copper in $CuSO_4$ is the oxidizing agent.
E. Sulfur is the reducing agent while oxygen is the oxidizing agent.
F. None of the above.

Question 180:

Which of the following statements is true?

A. Acids are compounds that act as proton acceptors in aqueous solution.
B. Acids only exist in a liquid state.
C. Strong acids are partially ionized in a solution.
D. Weak acids generally have a pH or 6 - 7.
E. The reaction between a weak and strong acid produces water and salt.

Question 181:

An unknown element, Z, has 3 isotopes: Z^5, Z^6 and Z^8. Given that the atomic mass of Z is 7, and the relative abundance of Z^5 is 20%, which of the following statements are correct?

1. Z^5 and Z^6 are present in the same abundance.
2. Z^8 is the most abundant of the isotopes.
3. Z^8 is more abundant than Z^5 and Z^6 combined.

A. 1 only
B. 2 only
C. 3 only
D. 1 and 2 only

E. 2 and 3 only
F. 1 and 3 only
G. 1, 2 and 3
H. None of the statements are correct.

Question 182:

Which of following best describes the products when an acid reacts with a metal that is more reactive than hydrogen?

A. Salt and hydrogen
B. Salt and ammonia
C. Salt and water
D. A weak acid and a weak base
E. A strong acid and a strong base
F. No reaction would occur.

Question 183:

Choose the option which balances the following equation:

$$\mathbf{a}\ FeSO_4 + \mathbf{b}\ K_2Cr_2O_7 + \mathbf{c}\ H_2SO_4 \rightarrow \mathbf{d}\ (Fe)_2(SO_4)_3 + \mathbf{e}\ Cr_2(SO_4)_3 + \mathbf{f}\ K_2SO_4 + \mathbf{g}\ H_2O$$

	a	b	c	d	e	f	g
A	6	1	8	3	1	1	7
B	6	1	7	3	1	1	7
C	2	1	6	2	1	1	6
D	12	1	14	4	1	1	14
E	4	1	12	4	1	1	12
F	8	1	8	4	2	1	8

Question 184:

Which of the following statements is correct?

A. Matter consists of atoms that have a net electrical charge.
B. Atoms and ions of the same element have different numbers of protons and electrons but the same number of neutrons.
C. Over 80% of an atom's mass is provided by protons.
D. Atoms of the same element that have different numbers of neutrons react at significantly different rates.
E. Protons in the nucleus of atoms repel each other as they are positively charged.
F. None of the above.

Question 185:

Which of the following statements is correct?

A. The noble gasses are chemically inert and therefore useless to man.
B. All the noble gasses have a full outer electron shell.
C. The majority of noble gasses are brightly coloured.
D. The boiling point of the noble gasses decreases as you progress down the group.
E. Neon is the most abundant noble gas.

Question 186:

In relation to alkenes, which of the following statements is correct?

1. They all contain double bonds.
2. They can all be reduced to alkanes.
3. Aromatic compounds are also alkenes as they contain double bonds.

A. Only 1
B. Only 2
C. Only 3
D. 1 and 2

E. 2 and 3
F. 1 and 3
G. All of the above.
H. None of the above.

Question 187:

Chlorine is made up of two isotopes, Cl^{35} (atomic mass 34.969) and Cl^{37} (atomic mass 36.966). Given that the atomic mass of chlorine is 35.453, which of the following statements is correct?

A. Cl^{35} is about 3 times more abundant than Cl^{37}.
B. Cl^{35} is about 10 times more abundant than Cl^{37}.
C. Cl^{37} is about 3 times more abundant than Cl^{35}.
D. Cl^{37} is about 10 times more abundant than Cl^{35}.
E. Both isotopes are equally abundant.

Question 188:
Which of the following statements regarding transition metals is correct?

A. Transition metals form ions that have multiple colours.
B. Transition metals usually form covalent bonds.
C. Transition metals cannot be used as catalysts as they are too reactive.
D. Transition metals are poor conductors of electricity.
E. Transition metals are frequently referred to as f-block elements.

Question 189:
20 g of impure Na^{23} reacts completely with excess water to produce 8,000 cm^3 of hydrogen gas under standard conditions. What is the percentage purity of sodium?
[Under standard conditions 1 mole of gas occupies 24 dm^3]

A. 88.0%   B. 76.5%   C. 66.0%   D. 38.0%   E. 15.3%

Question 190:
An organic molecule contains 70.6% Carbon, 5.9% Hydrogen and 23.5% Oxygen. It has a molecular mass of 136. What is its chemical formula?

A. C_4H_4O B. C_5H_4O C. $C_8H_8O_2$ D. $C_{10}H_8O_2$ E. C_2H_2O

Question 191:
Choose the option which balances the following reaction:

$$aS + bHNO_3 \rightarrow cH_2SO_4 + dNO_2 + eH_2O$$

	a	b	c	d	e
A	3	5	3	5	1
B	1	6	1	6	2
C	6	14	6	14	2
D	2	4	2	4	4
E	2	3	2	3	2
F	4	4	4	4	2

Question 192:
Which of the following statements is true?
1. Ethane and ethene can both dissolve in organic solvents.
2. Ethane and ethene can both be hydrogenated in the presence of Nickel.
3. Breaking C=C requires double the energy needed to break C-C.

A. 1 only
B. 2 only
C. 3 only
D. 1 and 2 only
E. 2 and 3 only
F. 1 and 3 only
G. 1, 2 and 3

Question 193:

Diamond, Graphite, Methane and Ammonia all exhibit covalent bonding. Which row adequately describes the properties associated with each?

	Compound	Melting Point	Able to conduct electricity	Soluble in water
1.	Diamond	High	Yes	No
2.	Graphite	High	Yes	No
3.	$CH_{4 (g)}$	Low	No	No
4.	$NH_{3 (g)}$	Low	No	Yes

A. 1 and 2 only   E. 1, 2 and 3

B. 2 and 3 only   F. 2, 3 and 4

C. 1 and 3 only   G. 1,2 and 4

D. 1 and 4 only   H. 1, 2, 3 and 4

Question 194:

Which of the following statements about catalysts are true?

1. Catalysts reduce the energy required for a reaction to take place.
2. Catalysts are used up in reactions.
3. Catalysed reactions are almost always exothermic.

A. 1 only   B. 2 only   C. 1 and 2   D. 2 and 3   E. 1, 2 and

Question 195:

What is the name of the molecule below?

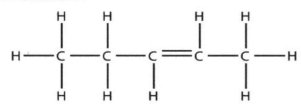

A. But-1-ene
B. But-2-ene
C. Pent-3-ene
D. Pent-1-ene
E. Pent-2-ene
F. Pentane
G. Pentanoic acid

Question 196:

Which of the following statements is correct regarding Group 1 elements? [Excluding Hydrogen]

A. The oxidation number of Group 1 elements usually decreases in most reactions.
B. Reactivity decreases as you progress down Group 1.
C. Group 1 elements do not react with water.
D. All Group 1 elements react instantaneously with oxygen.
E. All of the above.
F. None of the above.

Question 197:
Which of the following statements about electrolysis are correct?

1. The cathode attracts negatively charged ions.
2. Atoms are reduced at the anode.
3. Electrolysis can be used to separate mixtures.

A. Only 1
B. Only 2
C. Only 3
D. 1 and 2
E. 2 and 3
F. 1 and 3
G. 1, 2 and 3
H. None of the statement are correct.

Question 198:
Which of the following is **NOT** an isomer of pentane?

A. $CH_3CH_2CH_2CH_2CH_3$
B. $CH_3C(CH_3)CH_3CH_3$

C. $CH_3(CH_2)_3CH_3$
D. $CH_3C(CH_3)_2CH_3$

Question 199:
Choose the option which balances the following reaction:

$Cu + HNO_3 \rightarrow Cu(NO_3)_2 + NO + H_2O$

A. $8\ Cu + 3\ HNO_3 \rightarrow 8\ Cu(NO_3)_2 + 4\ NO + 2\ H_2O$
B. $3\ Cu + 8\ HNO_3 \rightarrow 2\ Cu(NO_3)_2 + 3\ NO + 4\ H_2O$
C. $5Cu + 7HNO_3 \rightarrow 5\ Cu(NO_3)_2 + 4\ NO + 8\ H_2O$
D. $6\ Cu + 10\ HNO_3 \rightarrow 6\ Cu(NO_3)_2 + 3\ NO + 7\ H_2O$
E. $3\ Cu + 8\ HNO_3 \rightarrow 3\ Cu(NO_3)_2 + 2\ NO + 4\ H_2O$

Question 200:
What of the following statements regarding alkenes is correct?

A. Alkenes are an inorganic homologous series.
B. Alkenes always have three times as many hydrogen atoms as they do carbon atoms.
C. Bromine water changes from clear to brown in the presence of an alkene.
D. Alkenes are more reactive than alkanes because they are unsaturated.
E. Alkenes frequently take part in subtraction reactions.
F. All of the above.

Question 201:
Which of the following statements is correct regarding Group 17?

A. All Group 17 elements are electrophilic and therefore form negatively charged ions.
B. All Group 17 elements are gasses a room temperature.
C. The reaction between Sodium and Fluorine is less vigorous than Sodium and Iodine.
D. All Group 17 elements are non-coloured.
E. Some Group 17 elements are found naturally as unbonded atoms.
F. All of the above.
G. None of the above.

Question 202:

Why does the electrolysis of NaCl solution (brine) require the strict separation of the products of anode and cathode?

A. To prevent the preferential discharge of ions.
B. In order to prevent spontaneous combustion.
C. In order to prevent production of H_2.
D. In order to prevent the formation of HCl.
E. In order to avoid CO poisoning.
F. All of the above.

Question 203:

In relation to the electrolysis of brine (NaCl), which of the following statements are correct?

1. Electrolysis results in the production of hydrogen and chlorine gas.
2. Electrolysis results in the production of sodium hydroxide.
3. Hydrogen gas is released at the anode and chlorine gas is released at the cathode.

A. Only 1
B. Only 2
C. Only 3
D. 1 and 2

E. 1 and 3
F. 2 and 3
G. All of the above

Question 204:

Which of the following statements is correct?

A. Alkanes consist of multiple C-H bonds that are very weak.
B. An alkane with 14 hydrogen atoms is called Heptane.
C. All alkanes consist purely of hydrogen and carbon atoms.
D. Alkanes burn in excess oxygen to produce carbon monoxide and water.
E. Bromine water is decolourised in the presence of an alkane.
F. None of the above.

Question 205:

Which of the following statements are correct?

1. All alcohols contain a hydroxyl functional group.
2. Alcohols are highly soluble in water.
3. Alcohols are sometimes used a biofuels.

A. Only 1
B. Only 2
C. Only 3
D. 1 and 2
E. 2 and 3
F. 1 and 3
G. 1, 2 and 3

Question 206:
Which row of the table below is correct?

	Non-Reducible Hydrocarbon			Reducible Hydrocarbon		
A	C_nH_{2n}	$Br_{2(aq)}$ remains brown	Saturated	C_nH_{2n+2}	Turns $Br_{2(aq)}$ colourless	Unsaturated
B	C_nH_{2n+2}	Turns $Br_{2(aq)}$ colourless	Unsaturated	C_nH_{2n}	$Br_{2(aq)}$ remains brown	Saturated
C	C_nH_{2n}	$Br_{2(aq)}$ remains brown	Unsaturated	C_nH_{2n+2}	Turns $Br_{2(aq)}$ colourless	Saturated
D	C_nH_{2n+2}	Turns $Br_{2(aq)}$ colourless	Saturated	C_nH_{2n}	$Br_{2(aq)}$ remains brown	Unsaturated
E	C_nH_{2n+2}	$Br_{2(aq)}$ remains brown	Saturated	C_nH_{2n}	Turns $Br_{2(aq)}$ colourless	Unsaturated

Question 207:
How many grams of magnesium chloride are formed when 10 grams of magnesium oxide are dissolved in excess hydrochloric acid? Relative atomic masses: $Mg = 24$, $O = 16$, $H = 1$, $Cl = 35.5$

A. 10.00
B. 14.95

C. 20.00
D. 23.75

E. 47.55
F. More information needed

Question 208:
Pentadecane has the molecular formula $C_{15}H_{32}$. Which of the following statements is true?

A. Pentadecane has a lower boiling point than pentane.
B. Pentadecane is more flammable than pentane.
C. Pentadecane is more volatile than pentane.
D. Pentadecane is more viscous than pentane.
E. All of the above.
F. None of the above.

Question 209:
The rate of reaction is normally dependent upon:
1. The temperature.
2. The concentration of reactants.
3. The concentration of the catalyst.
4. The surface area of the catalyst.

A. 1 and 2
B. 2 and 3

C. 2, 3 and 4
D. 1, 3 and 4

E. 1, 2 and 3
F. 1, 2, 3 and 4

Question 210:
The equation below shows the complete combustion of a sample of unknown hydrocarbon in excess oxygen.

$$C_aH_b + O_2 \rightarrow cCO_2 + dH_2O$$

The product yielded 176 grams of CO_2 and 108 grams of H_2O. What is the most likely formula of the unknown hydrocarbon? Relative atomic masses: $H = 1$, $C = 12$, $O = 16$.

A. CH_4
B. CH_3

C. C_2H_6
D. C_3H_9

E. C_2H_4
F. C_4H_{10}

Question 211:

What type of reaction must ethanol undergo in order to be converted to ethylene oxide (C_2H_4O)?

A. Oxidation
B. Reduction
C. Dehydration
D. Hydration
E. Redox
F. All of the above

Question 212:

What values of a, b and c balance the equation below?

$$a\ Ba_3N_2 + 6H_2O \rightarrow b\ Ba(OH)_2 + c\ NH_3$$

	a	b	c
A	1	2	3
B	1	3	2
C	2	1	3
D	2	3	1
E	3	1	2
F	3	2	1

Question 213:

What values of a, b and c balance the equation below?

$$a\ FeS + 7O_2 \rightarrow b\ Fe_2O_3 + c\ SO_2$$

	a	b	c
A	3	2	2
B	2	4	1
C	3	1	5
D	4	1	3
E	4	2	4

Question 214:

Magnesium consists of 3 isotopes: Mg^{23}, Mg^{25}, and Mg^{26} which are found naturally in a ratio of 80:10:10. Calculate the relative atomic mass of magnesium.

A. 23.3   B. 23.4   C. 23.5   D. 23.6   E. 24.6   F. 25.2   G. 25.5

Question 215:

Consider the three reactions:
1. $Cl_2 + 2Br^- \rightarrow 2Cl^- + Br_2$
2. $Cu^{2+} + Mg \rightarrow Cu + Mg^{2+}$
3. $Fe_2O_3 + 3CO \rightarrow 2Fe + 3CO_2$

Which of the following statements are correct?

A. Cl_2 and Fe_2O_3 are reducing agents.
B. CO and Cu^{2+} are oxidising agents.
C. Br_2 is a stronger oxidising agent than Cl_2.
D. Mg is a stronger reducing agent than Cu.
E. All of the above.
F. None of the above.

Question 216:

Which row best describes the properties of NaCl?

	Melting Point	Solubility in Water	Conducts electricity?	
			As solid	**In solution**
A	High	Yes	Yes	Yes
B	High	No	Yes	No
C	High	Yes	No	Yes
D	High	No	No	No
E	Low	Yes	Yes	Yes
F	Low	No	Yes	No
G	Low	Yes	No	Yes
H	Low	No	No	No

Question 217:

80g of Sodium hydroxide reacts with excess zinc nitrate to produce zinc hydroxide. Calculate the mass of zinc hydroxide produced. Relative atomic mass: N = 14, Zn = 65, O = 16, Na = 23.

A. 49g
B. 95g
C. 99g
D. 100g
E. 198g
F. More information needed.

Question 218:

Which of the following statements is correct?

A. The reaction between all Group 1 metals and water is exothermic.
B. All Group 1 metals react with water to produce a metal hydroxide.
C. All Group 1 metals react with water to produce elemental hydrogen.
D. Sodium reacts less vigorously with water than Potassium.
E. All of the above.
F. None of the above.

Question 219:

Which of the following statements is correct?

A. NaCl can be separated using sieves.
B. CO_2 can be separated using electrolysis.
C. Dyes in a sample of ink can be separated using chromatography.
D. Oil and water can be separated using fractional distillation.
E. Methane and diesel can be separated using a separating funnel.
F. None of the above.

Question 220:

Which of the following statements about the reaction between caesium and fluoride are correct?

1. It is an exothermic reaction and therefore requires catalysts.
2. It results in the formation of a salt.
3. The addition of water will make the reaction safer.

A. Only 1
B. Only 2
C. Only 3
D. 1 and 2

E. 2 and 3
F. 1 and 3
G. All of the above.
H. None of the above.

Question 221:

Which of the following statements is generally true about stable isotopes?
1. The nucleus contains an equal number of neutrons and protons.
2. The nuclear charge is equal and opposite to the peripheral charge due to the orbiting electrons.
3. They can all undergo radioactive decay into more stable isotopes.

A. Only 1
B. Only 2
C. Only 3
D. 1 and 2

E. 2 and 3
F. 1 and 3
G. All of the above.
H. None of the above.

Question 222:

Why do most salts have very high melting temperatures?
A. Their surface is able to radiate away a significant portion of heat to their environment.
B. The ionic bonds holding them together are very strong.
C. The covalent bonds holding them together are very strong.
D. They tend to form large macromolecules as each salt molecule bonds with multiple other molecules.
E. All of the above.

Question 223:

A bottle of water contains 306ml of pure deionised water. How many protons are in the bottle from the water? Avogadro Constant = 6×10^{23}.

A. 1×10^{22} B. 1×10^{23} C. 1×10^{24} D. 1×10^{25} E. 1×10^{26}

Question 224:

On analysis, an organic substance is found to contain 41.4% Carbon, 55.2% Oxygen and 3.45% Hydrogen by mass. Which of the following could be the empirical formula of this substance?
A. $C_3O_3H_6$
B. $C_3O_3H_{12}$
C. $C_4O_2H_4$
D. $C_4O_4H_4$
E. $C_4O_2H_8$
F. More information needed.

Question 225:

A is a Group 2 element and B is a Group 17 element. Which row best describes what happens when A reacts with B?

	B is	Formula
A	Reduced	AB
B	Reduced	A_2B
C	Reduced	AB_2
D	Oxidised	AB
E	Oxidised	A_2B
F	Oxidised	AB_2

SECTION 1D: Biology

Thankfully, the biology questions tend to be fairly straightforward and require the least **amount of time. You should** be able to do the majority of these within the time limit (often far less). This **means that you should be aiming** to make up time in these questions. In the majority of cases – you'll either know **the answer or not i.e. they test** advanced recall so the trick is to ensure that there are no obvious gaps in your **knowledge.**

Before going onto to do the practice questions in this book, ensure you are comfortable **with the following commonly** tested topics:

- ➤ Structure of animal, plant and bacterial cells
- ➤ Osmosis, Diffusion and Active Transport
- ➤ Cell Division (mitosis + meiosis)
- ➤ Family pedigrees and Inheritance
- ➤ DNA structure and replication
- ➤ Gene Technology & Stem Cells
- ➤ Enzymes – Function, mechanism and examples of digestive enzymes
- ➤ Aerobic and Anaerobic Respiration

- ➤ The central vs. peripheral **nervous system**
- ➤ The respiratory cycle **including movement of** ribs and diaphragm
- ➤ The Cardiac Cycle
- ➤ Hormones
- ➤ Basic immunology
- ➤ Food chains and food **webs**
- ➤ The carbon and nitrogen **cycles**

Top tip! If you find yourself getting less than 50% of biology questions correct **in this book, make sure you** revisit the syllabus before attempting more questions as this is the best way to **maximise your efficiency. In** general, there is no reason why you shouldn't be able to get the vast majority of biology **questions correct (and** in well under 60 seconds) with sufficient practice.

Biology Questions

Question 226:

In relation to the human genome, which of the following are correct?

1. The DNA genome is coded by 4 different bases.
2. The sugar backbone of the DNA strand is formed of glucose.
3. DNA is found in the nucleus of bacteria.

A. 1 only
B. 2 only
C. 3 only
D. 1 and 2
E. 1 and 3
F. 2 and 3
G. 1, 2 and 3

Question 227:

Animal cells contain organelles that take part in vital processes. Which of the following is true?

1. The majority of energy production by animal cells occurs in the mitochondria.
2. The cell wall protects the animal cell membrane from outside pressure differences.
3. The endoplasmic reticulum plays a role in protein synthesis.

A. 1 only
B. 2 only
C. 3 only
D. 1 and 2
E. 2 and 3
F. 1 and 3
G. 1, 2 and 3

Question 228:

With regards to animal mitochondria, which of the following is correct?

A. Mitochondria are not necessary for aerobic respiration.
B. Mitochondria are the sole cause of sperm cell movement.
C. The majority of DNA replication happens inside mitochondria.
D. Mitochondria are more abundant in fat cells than in skeletal muscle.
E. The majority of protein synthesis occurs in mitochondria.
F. Mitochondria are enveloped by a double membrane.

Question 229:

In relation to bacteria, which of the following is **FALSE**?

A. Bacteria always lead to disease.
B. Bacteria contain plasmid DNA.
C. Bacteria do not contain mitochondria.
D. Bacteria have a cell wall and a plasma membrane.
E. Some bacteria are susceptible to antibiotics.

Question 230:

In relation to bacterial replication, which of the following is correct?

A. Bacteria undergo sexual reproduction.
B. Bacteria have a nucleus.
C. Bacteria carry genetic information on circular plasmids.
D. Bacterial genomes are formed of RNA instead of DNA.
E. Bacteria require gametes to replicate.

Question 231:

Which of the following are correct regarding active transport?

A. ATP is necessary and sufficient for active transport.
B. ATP is not necessary but sufficient for active transport.
C. The relative concentrations of the material being transported have little impact on the rate of active transport.
D. Transport proteins are necessary and sufficient for active transport.
E. Active transport relies on transport proteins that are powered by an electrochemical gradient.

Question 232:

Concerning mammalian reproduction, which of the following is **FALSE**?

A. Fertilisation involves the fusion of two gametes.
B. Reproduction is sexual and the offspring display genetic variation.
C. Reproduction relies upon the exchange of genetic material.
D. Mammalian gametes are diploid cells produced via meiosis.
E. Embryonic growth requires carefully controlled mitosis.

Question 233:

Which of the following apply to Mendelian inheritance?

1. It only applies to plants.
2. It treats different traits as either dominant or recessive.
3. Heterozygotes have a 25% chance of expressing a recessive trait.

A. 1 only
B. 2 only
C. 3 only
D. 1 and 2
E. 1 and 3
F. 2 and 3
G. All of the above.

Question 234:

Which of the following statements are correct?

A. Hormones are secreted into the blood stream and act over long distances at specific target organs.
B. Hormones are substances that almost always cause muscles to contract.
C. Hormones have no impact on the nervous or enteric systems.
D. Hormones are always derived from food and never synthesised.
E. Hormones act rapidly to restore homeostasis.

Question 235:

With regard to neuronal signalling in the body, which of the following are true?

1. Neuronal transmission can be caused by both electrical and chemical stimulation.
2. Synapses ultimately result in the production of an electrical current for signal transduction.
3. All synapses in humans are electrical and unidirectional.

A. 1 only
B. 2 only
C. 3 only
D. 1 and 2
E. 1 and 3
F. 2 and 3
G. 1, 2 and 3

Question 236:

What is the **primary** reason that pH is controlled so tightly in humans?

A. To allow rapid protein synthesis.
B. To allow for effective digestion throughout the GI tract.
C. To ensure ions can function properly in neural signalling.
D. To prevent changes in electrical charge in polypeptide chains.
E. To prevent changes in core body temperature.

Question 237:

Which of the following statements are correct regarding cell walls?

1. The cell wall confers protection against external environmental stimuli.
2. The cell wall is an evolutionary remnant and now has little functional significance in most bacteria.
3. The cell wall is made up primarily of glucose.

A. Only 1
B. Only 2
C. Only 3
D. 1 and 2
E. 2 and 3
F. 1 and 3
G. 1, 2 and 3

Question 238:

Which of the following statements are correct regarding mitosis?

1. It is important in sexual reproduction.
2. A single round of mitosis results in the formation of 2 genetically distinct daughter cells.
3. Mitosis is vital for tissue growth, as it is the basis for cell multiplication.

A. Only 1
B. Only 2
C. Only 3
D. 1 and 2
E. 2 and 3
F. 1 and 3
G. 1, 2 and 3

Question 239:

Which of the following is the best definition of a mutation?

A. A mutation is a permanent change in DNA.
B. A mutation is a permanent change in DNA that is harmful to an organism.
C. A mutation is a permanent change in the structure of intra-cellular organelles caused by changes in DNA/RNA.
D. A mutation is a permanent change in chromosomal structure caused by DNA/RNA changes.

Question 240:

In relation to mutations, which of the following are correct?

1. Mutations always lead to discernible changes in the phenotype of an organism.
2. Mutations are central to natural processes such as evolution.
3. Mutations play a role in cancer.

A. Only 1
B. Only 2
C. Only 3
D. 1 and 2
E. 2 and 3
F. 1 and 3
G. 1, 2 and 3

Question 241:

Which of the following is the most accurate definition of an antibody?

A. An antibody is a molecule that protects red blood cells from changes in pH.
B. An antibody is a molecule produced only by humans and has a pivotal role in the immune system.
C. An antibody is a toxin produced by a pathogen to damage the host organism.
D. An antibody is a molecule that is used by the immune system to identify and neutralize foreign objects and molecules.
E. Antibodies are small proteins found in red blood cells that help increase oxygen carriage.

Question 242:

Which of the following statements about the kidney are correct?

1. The kidneys filter the blood and remove waste products from the body.
2. The kidneys are involved in the digestion of food.
3. In a healthy individual, the kidneys produce urine that contains high levels of glucose.

A. Only 1
B. Only 2
C. Only 3
D. 1 and 2
E. 2 and 3
F. 1 and 3
G. 1, 2 and 3

Question 243:
Which of the following statements are correct?

1. Hormones are slower acting than nerves.
2. Hormones act for a very short time.
3. Hormones act more generally than nerves.
4. Hormones are released when you get a scare.

A. 1 only
B. 1 and 3 only
C. 2 and 4 only
D. 1, 3 and 4 only
E. 1, 2, 3 and 4

Question 244:
Which statements about homeostasis are correct?

1. Homeostasis is about ensuring the inputs within your body exceed the outputs to maintain a constant internal environment.
2. Homeostasis is about ensuring the inputs within your body are less than the outputs to maintain a constant internal environment.
3. Homeostasis is about balancing the inputs within your body with the outputs to ensure your body fluctuates with the needs of the external environment.
4. Homeostasis is about balancing the inputs within your body with the outputs to maintain a constant internal environment.

A. 1 only
B. 2 only
C. 3 only
D. 4 only
E. 1 and 3 only
F. 2 and 4 only
G. 2 and 3 only

Question 245:
Which of the following statement is true?

A. There is more energy and biomass each time you move up a trophic level.
B. There is less energy and biomass each time you move up a trophic level.
C. There is more energy but less biomass each time you move up a trophic level.
D. There is less energy but more biomass each time you move up a trophic level.
E. There is no difference in the energy or biomass when you move up a trophic level.

Question 246:
Which of the following statements are true about asexual reproduction?

1. There is no fusion of gametes.
2. There are two parents.
3. There is no mixing of chromosomes.
4. There is genetic variation.

A. 1 and 3 only
B. 1 and 4 only
C. 2 and 3 only
D. 3 and 4 only
E. 2 and 4 only
F. 1, 2, 3 and 4

Question 247:
Put the following in the order which they occur when Jonas sees a bowl of chicken and moves towards it.

1. Retina
2. Motor neuron
3. Sensory neuron
4. Brain
5. Muscle

A. 1 - 3 - 4 - 5 - 2
B. 1 - 2 - 3 - 4 - 5
C. 5 - 1 - 3 - 2 - 4

D. 1 - 3 - 2 - 4 - 5
E. 1 - 3 - 4 - 2 - 5
F. 4 - 1 - 3 - 2 - 5

Question 248:
What path does blood take from the kidney to the liver?

1. Pulmonary artery
2. Inferior vena cava
3. Hepatic artery
4. Aorta
5. Pulmonary vein
6. Renal vein

A. 2 - 1 - 4 - 3 - 5 - 6
B. 1 - 2 - 3 - 4 - 5 - 6
C. 6 - 2 - 5 - 1 - 4 - 3

D. 6 - 2 - 1 - 5 - 4 - 3
E. 3 - 2 - 1 - 4 - 6 - 5
F. 3 - 6 - 2 - 4 - 1 – 5

Question 249:
Which of the following statements are true about animal cloning?

1. Animals cloned from embryo transplants are genetically identical.
2. The genetic material is removed from an unfertilised egg during adult cell cloning.
3. Cloning can cause a reduced gene pool.
4. Cloning is only possible with mammals.

A. 1 only
B. 2 only
C. 3 only
D. 4 only
E. 1 and 2 only
F. 1, 2 and 3 only
G. 1, 2, 3 and 4

Question 250:
Which of the following statements are true with regard to evolution?

1. Individuals within a species show variation because of differences in their genes.
2. Beneficial mutations will accumulate within a population.
3. Gene differences are caused by sexual reproduction and mutations.
4. Species with similar characteristics never have similar genes.

A. 1 only
B. 1 and 4 only

C. 2 and 3 only
D. 2 and 4 only

E. 3 and 4 only
F. 1, 2 and 3 only

Question 251:
Which of the following genetic statements are correct?

1. Alleles are a similar version of different cells.
2. If you are homozygous for a trait, you have three alleles the same for that particular gene.
3. If you are heterozygous for a trait, you have two different alleles for that particular gene.
4. To show the characteristic that is caused by a recessive allele, both carried alleles for the gene have to be recessive.

A. 1 only
B. 2 only
C. 3 only
D. 4 only
E. 1 and 2 only
F. 3 and 4 only
G. 1, 2, and 3 only

Question 252:
Which of the following statements are correct about meiosis?

1. The DNA content of a gamete is half that of a human red blood cell.
2. Meiosis requires ATP.
3. Meiosis only takes place in reproductive tissue.
4. In meiosis, a diploid cell divides in such a way so as to produce two haploid cells.

A. 1 only
B. 3 only
C. 1 and 2 only
D. 2 and 3 only
E. 2 and 4 only
F. 1, 2, 3 and 4

Question 253:
Put the following statements in the correct order of events for when there is too little water in the blood.

1. Urine is more concentrated
2. Pituary gland releases ADH
3. Blood water level returns to normal
4. Hypothalamus detects too little water in blood
5. Kidney affects water level

A. 1 - 2 - 3 - 4 - 5
B. 5 - 4 - 3 - 2 - 1
C. 4 - 2 - 5 - 1 - 3
D. 3 - 2 - 4 - 1 - 5
E. 5 - 2 - 3 - 4 - 1
F. 4 - 2 – 1- 5 - 3

Question 254:
The pH of venous blood is 7.35. Which of the following is the likely pH of arterial blood?

A. 4.4   C. 6.5   E. 7.4
B. 5.2   D. 7.0   F. 7.95

Question 255:

Which of the following are true of the cytoplasm?

1. The vast majority of the cytoplasm is made up of water.
2. All contents of animal cells are contained in the cytoplasm.
3. The cytoplasm contains electrolytes and proteins.

A. 1 only
B. 2 only
C. 3 only
D. 1 and 2 only
E. 1 and 3 only
F. 1, 2 and 3

Question 256:

ATP is produced in which of the following organelles?

1. The golgi apparatus
2. The rough endoplasmic reticulum
3. The mitochondria
4. The nucleus

A. 1 only
B. 2 only
C. 3 only
D. 4 only
E. 1 and 2
F. 2 and 3 only
G. 3 and 4 only
H. 1, 2, 3 and 4

Question 257:

The cell membrane:
A. Is made up of a phospholipid bilayer which only allows active transport across it.
B. Is not found in bacteria.
C. Is a semi-permeable barrier to ions and organic molecules.
D. Consists purely of enzymes.

Question 258:

Cells of the *Polyommatus atlantica* butterfly of the Lycaenidae family have 446 chromosomes. Which of the following statements about a *P. atlantica* butterfly are correct?

1. Mitosis will produce 2 daughter cells each with 223 pairs of chromosomes
2. Meiosis will produce 4 daughter cells each with 223 chromosomes
3. Mitosis will produce 4 daughter cells each with 446 chromosomes
4. Meiosis will produce 2 daughter cells each with 223 pairs of chromosomes

A. 1 and 2 only
B. 1 and 3 only
C. 2 and 3 only
D. 3 and 4 only
E. 1, 2 and 3 only
F. 1, 2, 3 and 4

71

Questions 259-261 are based on the following information:

Assume that hair colour is determined by a single allele. The R allele is dominant and results in black hair. The r allele is recessive for red hair. Mary (red hair) and Bob (black hair) are having a baby girl.

Question 259:

What is the probability that she will have red hair?

A. 0% only
B. 25% only
C. 50% only
D. 0% or 25%
E. 0% or 50%
F. 25% or 50%

Question 260:

Mary and Bob have a second child, Tim, who is born with red hair. What does this confirm about Bob?

A. Bob is heterozygous for the hair allele.
B. Bob is homozygous dominant for the hair allele.
C. Bob is homozygous recessive for the hair allele.
D. Bob does not have the hair allele.

Question 261:

Mary and Bob go on to have a third child. What are the chances that this child will be born homozygous for black hair?

A. 0%
B. 25%
C. 50%
D. 75%
E. 100%

Question 262:

Why does air flow into the chest on inspiration?

1. Atmospheric pressure is smaller than intra-thoracic pressure during inspiration.
2. Atmospheric pressure is greater than intra-thoracic pressure during inspiration.
3. Anterior and lateral chest expansion decreases absolute intra-thoracic pressure.
4. Anterior and lateral chest expansion increases absolute intra-thoracic pressure.

A. 1 only
B. 2 only
C. 2 and 3
D. 1 and 4
E. 1 and 3
F. 2 and 4

Question 263:
Which of the following components of a food chain represent the largest biomass?

A. Producers
B. Decomposers
C. Primary consumers
D. Secondary consumers
E. Tertiary consumers

Question 264:
Concerning the nitrogen cycle, which of the following are true?

1. The majority of the Earth's atmosphere is nitrogen.
2. Most of the nitrogen in the Earth's atmosphere is inert.
3. Bacteria are essential for nitrogen fixation.
4. Nitrogen fixation occurs during lightning strikes.

A. 1 and 2
B. 1 and 3
C. 2 and 3
D. 2 and 4
E. 3 and 4
F. 1, 2, 3 and 4

Question 265:
Which of the following statement are correct regarding mutations?

1. Mutations always cause proteins to lose their function.
2. Mutations always change the structure of the protein encoded by the affected gene.
3. Mutations always result in cancer.

A. Only 1
B. Only 2
C. Only 3
D. 1 and 2
E. 2 and 3
F. 1 and 3
G. 1, 2 and 3
H. None of the statements are correct.

Question 266:
Which of the following is not a function of the central nervous system?

A. Coordination of movement
B. Decision making and executive functions
C. Control of heart rate
D. Cognition
E. Memory

Question 267:
Which of the following control mechanisms are involved in modulating cardiac output?

1. Voluntary control.
2. Sympathetic control to decrease heart rate.
3. Parasympathetic control to increase heart rate.

A. Only 1
B. Only 2
C. Only 3
D. 1 and 2
E. 2 and 3
F. 1 and 3
G. 1, 2 and 3
H. None of the statements are correct.

Question 268:
Vijay goes to see his GP with fatty, smelly stools that float on water. Which of the following enzymes is most likely to be malfunctioning?

A. Amylase
B. Lipase
C. Protease
D. Sucrase
E. Lactase

Question 269:
Which of the following statements concerning the cardiovascular system is correct?

A. Oxygenated blood from the lungs flows to the heart via the pulmonary artery.
B. All arteries carry oxygenated blood.
C. All animals have a double circulatory system.
D. The superior vena cava contains oxygenated blood
E. All veins have valves.
F. None of the above.

Question 270:
Which part of the GI tract has the least amount of enzymatic digestion occurring?

A. Mouth
B. Stomach
C. Small intestine
D. Large intestine
E. Rectum

Question 271:
Oge touches a hot stove and immediately moves her hand away. Which of the following **components are NOT involved** in this reaction?

1. Thermo-receptor
2. Brain
3. Spinal Cord
4. Sensory nerve
5. Motor nerve
6. Muscle

A. 1 only
B. 2 only
C. 3 only
D. 1 and 2 only
E. 1, 2 and 3 only
F. 3, 4, 5 and 6

Question 272:
Which of the following represents a scenario with an appropriate description of the **mode of transport?**

1. Water moving from a hypotonic solution outside of a potato cell, across the cell **wall and cell membrane and** into the hypertonic cytoplasm of the potato cell→ Osmosis.
2. Carbon dioxide moving across a respiring cell's membrane and dissolving in **blood plasma** →Active transport.
3. Reabsorption of amino acids against a concentration gradient in the glomeruluar **apparatus** → Diffusion.

A. 1 only
B. 2 only
C. 3 only
D. 1 and 2 only
E. 2 and 3 only
F. 1 and 3 only
G. 1, 2 and 3

Question 273:
Which of the following equations represents anaerobic respiration?

1. Carbohydrate + Oxygen $\rightarrow$ Energy + Carbon Dioxide + Water
2. Carbohydrate $\rightarrow$ Energy + Lactic Acid + Carbon dioxide
3. Carbohydrate $\rightarrow$ Energy + Lactic Acid
4. Carbohydrate $\rightarrow$ Energy + Ethanol + Carbon dioxide

A. 1 only
B. 2 only
C. 3 only
D. 4 only
E. 1 and 2
F. 1 and 3
G. 1 and 4
H. 2 and 4 only
I. 3 and 4 only

Question 274:
Which of the following statements regarding respiration are correct?

1. The mitochondria are the centres for both aerobic and anaerobic respiration.
2. The cytoplasm is the main site of anaerobic respiration.
3. For every two moles of glucose that is respired aerobically, 12 moles of CO_2 are liberated.
4. Anaerobic respiration is more efficient than aerobic respiration.

A. 1 and 2
B. 1 and 4
C. 2 and 3
D. 2 and 4
E. 3 and 4

Question 275:
Which of the following statements are true?

1. The nucleus contains the cell's chromosomes.
2. The cytoplasm consists purely of water.
3. The plasma membrane is a single phospholipid layer.
4. The cell wall prevents plants cells from lysing due to osmotic pressure.

A. 1 and 2
B. 1 and 4
C. 1, 3 and 4
D. 1, 2 and 3
E. 1, 2 and 4
F. 2, 3 and 4

Question 276:
Which of the following statements are true about osmosis?

1. If a medium is hypertonic relative to the cell cytoplasm, the cell will gain water through osmosis.
2. If a medium is hypotonic relative to the cell cytoplasm, the cell will gain water through osmosis.
3. If a medium is hypotonic relative to the cell cytoplasm, the cell will lose water through osmosis.
4. If a medium is hypertonic relative to the cell cytoplasm, the cell will lose water through osmosis.
5. The medium's tonicity has no impact on the movement of water.

A. 1 only
B. 2 only
C. 1 and 3
D. 2 and 4
E. 5 only

Question 277:
Which of the following statements are true about stem cells?

1. Stem cells have the ability to differentiate into other mature types of cells.
2. Stem cells are unable to maintain their undifferentiated state.
3. Stem cells can be classified as embryonic stem cells or adult stem cells.
4. Stem cells are only found in embryos.

A. 1 and 3
B. 3 and 4
C. 2 and 3
D. 1 and 2
E. 2 and 4

Question 278:
Which of the following are **NOT** examples of natural selection?

1. Giraffes growing longer necks to eat taller plants.
2. Antibiotic resistance developed by certain strains of bacteria.
3. Pesticide resistance among locusts in farms.
4. Breeding of horses to make them run faster.

A. 1 only
B. 4 only
C. 1 and 3
D. 1 and 4
E. 2 and 4

Question 279:
Which of the following statements are true?

1. Enzymes stabilise the transition state and therefore lower the activation energy.
2. Enzymes distort substrates in order to lower activation energy.
3. Enzymes decrease temperature to slow down reactions and lower the activation energy.
4. Enzymes provide alternative pathways for reactions to occur.

A. 1 only
B. 1 and 2
C. 1 and 4
D. 2 and 4
E. 3 and 4

Question 280:

Which of the following are examples of negative feedback?

1. Salivating whilst waiting for a meal.
2. Throwing a dart.
3. The regulation of blood pH.
4. The regulation of blood pressure.

A. 1 only
B. 1 and 2
C. 3 and 4
D. 2, 3, and 4
E. 1, 2, 3 and 4

Question 281:

Which of the following statements about the immune system are true?

1. White blood cells defend against bacterial and fungal infections.
2. White blood cells can temporarily disable but not kill pathogens.
3. White blood cells use antibodies to fight pathogens.
4. Antibodies are produced by bone marrow stem cells.

A. 1 and 3
B. 1 and 4
C. 2 and 3
D. 2 and 4
E. 1, 2, and 3
F. 1, 3, and 4

Question 282:

The cardiovascular system does **NOT**:

A. Deliver vital nutrients to peripheral cells.
B. Oxygenate blood and transports it to peripheral cells.
C. Act as a mode of transportation for hormones to reach their target organ.
D. Facilitate thermoregulation.
E. Respond to exercise by increasing cardiac output to exercising muscles.

Question 283:

Which of the following statements is correct?

A. Adrenaline can sometimes decrease heart rate.
B. Adrenaline is rarely released during flight or fight responses.
C. Adrenaline causes peripheral vasoconstriction.
D. Adrenaline only affects the cardiovascular system.
E. Adrenaline travels primarily in lymphatic vessels.
F. None of the above.

Question 284:
Which of the following statements is true?

A. Protein synthesis occurs solely in the nucleus.
B. Each amino acid is coded for by three DNA bases.
C. Each protein is coded for by three amino acids.
D. Red blood cells can create new proteins to prolong their lifespan.
E. Protein synthesis isn't necessary for mitosis to take place.
F. None of the above.

Question 285:
A solution of amylase and carbohydrate is present in a beaker, where the pH of the contents is 6.3. Assuming amylase is saturated, which of the following will increase the rate of production of the product?

1. Add sodium bicarbonate
2. Add carbohydrate
3. Add amylase
4. Increase the temperature to 100° C

A. 1 only   C. 3 only   E. 1 and 2   G. 1, 2 and 3
B. 2 only   D. 4 only   F. 1 and 3   H. 1, 3 and 4

Question 286:
Celestial Necrosis is a newly discovered autosomal recessive disorder. A female carrier and a male with the disease produce two boys. What is the probability that neither boy's genotype contains the celestial necrosis allele?

A. 100%   B. 75%   C. 50%   D. 25%   E. 0%

Question 287:
Which among the following has no endocrine function?

A. The thyroid   C. The pancreas   E. The testes
B. The ovary   D. The adrenal gland   F. None of the above.

Question 288:
Which of the following statements are true?

1. Increasing levels of insulin cause a decrease in blood glucose levels.
2. Increasing levels of glycogen cause an increase in blood glucose levels.
3. Increasing levels of adrenaline decrease the heart rate.

A. 1 only
B. 2 only
C. 3 only
D. 1 and 2
E. 2 and 3
F. 1 and 3
G. 1, 2 and 3

Question 289:

Which of the following rows is correct?

	Oxygenated Blood		Deoxygenated Blood	
A.	Left atrium	Left ventricle	Right atrium	Right ventricle
B.	Left atrium	Right atrium	Left ventricle	Right ventricle
C.	Left atrium	Right ventricle	Right atrium	Right ventricle
D.	Right atrium	Right ventricle	Left atrium	Left ventricle
E.	Left ventricle	Right atrium	Left atrium	Right ventricle

Questions 290-292 are based on the following information:

The pedigree below shows the inheritance of a newly discovered disease that affects connective tissue called Nafram syndrome. Individual 1 is a normal homozygote.

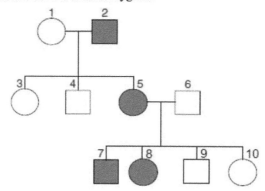

Question 290:

What is the inheritance of Nafram syndrome?

A. Autosomal dominant
B. Autosomal recessive
C. X-linked dominant

D. X-linked recessive
E. Co-dominant

Question 291:

Which individuals must be heterozygous for Nafram syndrome?

A. 1 and 2
B. 8 and 9

C. 2 and 5
D. 5 and 6

E. 6 and 8
F. 6 and 10

Question 292:

Taking N to denote a diseased allele and n to denote a normal allele, which of the following are **NOT** possible genotypes for 6's parents?

1. NN x NN
2. NN x Nn
3. Nn x nn
4. Nn x Nn
5. nn x nn

A. 1 and 2
B. 1 and 3

C. 2 and 3
D. 2 and 5

E. 3 and 4
F. 4 and 5

Question 293:

Which of the following correctly describes the passage of urine through the body?

	1st	2nd	3rd	4th
A	Kidney	Ureter	Bladder	Urethra
B	Kidney	Urethra	Bladder	Ureter
C	Urethra	Bladder	Ureter	Kidney
D	Ureter	Kidney	Bladder	Urethra

Question 294:

Which of the following best describes the passage of blood from the body, through the heart, back to the body?

A. Aorta → Left Ventricle → Left Atrium → Inferior Vena Cava → Right Atrium → Right Ventricle → Lungs → Aorta

B. Inferior vena cava → Left Atrium → Left Ventricle → Lungs → Right Atrium → Right Ventricle → Aorta

C. Inferior vena cava → Right Ventricle → Right Atrium → Lungs → Left Atrium → Left Ventricle → Aorta

D. Aorta → Left Atrium → Left Ventricle → Lungs → Right Atrium → Right Ventricle → Inferior Vena Cava

E. Right Atrium → Left Atrium → Inferior vena cava → Lungs → Left Atrium → Right Ventricle → Aorta

F. None of the above.

Question 295:

Which of the following best describes the events during inspiration?

	Intrathoracic Pressure	Intercostal Muscles	Diaphragm
A	Increases	Contract	Contracts
B	Increases	Relax	Contracts
C	Increases	Contract	Relaxes
D	Increases	Relax	Relaxes
E	Decreases	Contract	Contracts
F	Decreases	Relax	Contracts
G	Decreases	Contract	Relaxes
H	Decreases	Relax	Relaxes

Questions 296-297 are based on the following information:

DNA is made up of the four nucleotide bases: adenine, cytosine, guanine and thymine. A triplet repeat or codon is a sequence of three nucleotides which code for an amino acid. While there are only 20 amino acids there are 64 different combinations of the four DNA nucleotide bases. This means that more than one combination of 3 DNA nucleotides sequences code for the same amino acid.

Question 296:
Which property of the DNA code is described above?

A. The code is unambiguous.
B. The code is universal.
C. The code is non-overlapping.
D. The code is degenerate.
E. The code is preserved.
F. The code has no punctuation.

Question 297:
Which type of mutation does the described property protect against the most?
A. An insertion - where a single nucleotide is inserted.
B. A point mutation - where a single nucleotide is replaced for another.
C. A deletion - where a single nucleotide is deleted.
D. A repeat expansion - where a repeated trinucleotide sequence is added.
E. A duplication - where a piece of DNA is abnormally copied.

Question 298:
Which row of the table below describes what happens when external temperature decreases?

	Temperature Change Detected by	Sweat Gland Secretion	Cutaneous Blood Flow
A	Hypothalamus	Increases	Increases
B	Hypothalamus	Increases	Decreases
C	Hypothalamus	Decreases	Increases
D	Hypothalamus	Decreases	Decreases
E	Cerebral Cortex	Increases	Increases
F	Cerebral Cortex	Increases	Decreases
G	Cerebral Cortex	Decreases	Increases
H	Cerebral Cortex	Decreases	Decreases

Question 299:
Which of the following processes involve active transport?

1. Reabsorption of glucose in the kidney.
2. Movement of carbon dioxide into the alveoli in the lungs.
3. Movement of chemicals in a neural synapse.

A. 1 only
B. 2 only
C. 3 only
D. 1 and 2
E. 1 and 3
F. 2 and 3
G. 1, 2 and 3

Question 300:
Which of the following statements is correct about enzymes?

A. All enzymes are made up of amino acids only.
B. Enzymes can sometimes slow the rate of reactions.
C. Enzymes have no impact on reaction temperatures.
D. Enzymes are heat sensitive but resistant to changes in pH.
E. Enzymes are unspecific in their substrate use.
F. None of the above.

END OF SECTION

SECTION 1E: Advanced Maths

Section 1E requires a much broader knowledge of the A level Maths curriculum and you're highly advised to revise the topics below before proceeding further with the practice questions in this book.

Algebra:
➢ Laws of Indices
➢ Manipulation of Surds
➢ Quadratic Functions: Graphs, use of discriminant, completing the square
➢ Solving Simultaneous Equations via Substitution
➢ Solving Linear and Quadratic Inequalities
➢ Manipulation of polynomials e.g. expanding brackets, factorising
➢ Use of Factor Theorem + Remainder Theorem

Graphing Functions:
➢ Sketching of common functions including lines, quadratics, cubics, trigonometric functions, logarithmic functions and exponential functions
➢ Manipulation of functions using simple transformations
➢ Graph of $y = a^x$ series

Law of Logarithms
➢ $a^b = c \leftrightarrow b = log_a c$
➢ $log_a x + log_a y = log_a(xy)$
➢ $log_a x - log_a y = log_a(\frac{x}{y})$
➢ $k\, log_a x = log_a(x^k)$
➢ $log_a \frac{1}{x} = - log_a x$
➢ $log_a a = 1$

Differentiation:
➢ First order and second order derivatives
➢ Familiarity with notation: $\frac{dy}{dx}, \frac{d^2y}{dx^2}, f'(x), f''(x)$
➢ Differentiation of functions like $y = x^n$

Integration:
➢ Definite and indefinite integrals for $y = x^n$
➢ Solving Differential Equations in the form: $\frac{dy}{dx} = f(x)$
➢ Understanding of the Fundamental Theorem of Calculus and its application:
  o $\int_a^b f(x)dx = F(b) - F(a), where\ F'(x) = f(x)$
  o $\frac{d}{dx}\int_a^x f(t)dt = f(x)$

Logic Arguments:
➢ Terminology: True, false, and, or not, necessary, sufficient, for all, for some, there exists.
➢ Arguments in the format:
  o If A then B
  o A if B
  o A only if B
  o A if and only if B

Geometry:
- Equations for a circle:
  - $(x - a)^2 + (y - b)^2 = r^2$
  - $x^2 + y^2 + cx + dy + e = 0$
- Equations for a straight line: $y - y_1 = m(x - x_1)$ & $Ax + by + c = 0$
- Circle Properties:
  - The angle subtended by an arc at the centre of a circle is double the size of the angle subtended by the arc on the circumference
  - The opposite angles in a cyclic quadrilateral summate to 180 degrees
  - The angle between the tangent and chord at the point of contact is equal to the angle in the alternate segment
  - The tangent at any point on a circle is perpendicular to the radius at that point
  - Triangles formed using the full diameter are right-angled triangles
  - Angles in the same segment are equal
  - The Perpendicular from the centre to a chord bisects the chord

Series:
- Arithmetic series and Geometric Series
- Summing to a finite and infinite geometric series
- Binomial Expansions
- Factorials

Trigonometry:
- Solution of trigonometric identities
- Values of sin, cost, tan for 0, 30, 45, 60 and 90 degrees
- Sine, Cosine, Tangent graphs, symmetries, periodicities
- Sin Rule: $\frac{a}{SinA} = \frac{b}{Sin\,B} = \frac{c}{Sin\,C}$
- Cosine Rule: $c^2 = a^2 + b^2 - 2ab\,cosC$
- $Area\ of\ Triangle = \frac{1}{2}ab\sin C$
- $\sin^2 \theta + \cos^2 \theta = 1$
- $tan\theta = \frac{sin\theta}{cos\,\theta}$

Advanced Maths Questions

Question 301:
The vertex of an equilateral triangle is covered by a circle whose radius is half the height of the triangle. What percentage of the triangle is covered by the circle?

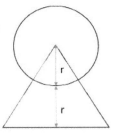

A. 12% C. 23% E. 41%
B. 16% D. 33% F. 50%

Question 302:
Three equal circles fit into a quadrilateral as shown, what is the height of the quadrilateral?

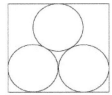

A. $2\sqrt{3}r$ D. $3r$
B. $(2 + \sqrt{3})r$ E. $4r$
C. $(4 - \sqrt{3})r$ F. More Information Needed

Question 303:
Two pyramids have equal volume and height, one with a square of side length a and one with a hexagonal base of side length b. What is the ratio of the side length of the bases?

A. $\sqrt{\frac{3\sqrt{3}}{2}}$ B. $\sqrt{\frac{2\sqrt{3}}{3}}$ C. $\sqrt{\frac{3}{2}}$ D. $\frac{2\sqrt{3}}{3}$ E. $\frac{3\sqrt{3}}{2}$

Question 304:
One 9 cm cube is cut into 3 cm cubes. The total surface area increases by a factor of:

A. $\frac{1}{3}$ B. $\sqrt{3}$ C. 3 E. 27
     D. 9

Question 305:
A cone has height twice its base width (four times the circle radius). What is the cone angle (half the angle at the vertex)?

A. 30° C. $\sin^{-1}\left(\frac{1}{\sqrt{17}}\right)$
B. $\sin^{-1}\left(\frac{r}{2}\right)$ D. $\cos^{-1}(\sqrt{17})$

Question 306:
A hemispherical speedometer has a maximum speed of 200 mph. What is the angle travelled by the needle at a speed of 70 mph?

A. 28° B. 49° C. 63° D. 88° E. 92°

Question 307:
Two rhombuses, A and B, are similar. The area of A is 10 times that of B. What is the ratio of the smallest angles over the ratio of the shortest sides?

A. 0 B. $\frac{1}{10}$ C. $\frac{1}{\sqrt{10}}$ D. $\sqrt{10}$ E. ∞

Question 308:

If $f^{-1}(-x) = \ln(2x^2)$ what is $f(x)$?

A. $\sqrt{\dfrac{e^y}{2}}$ 　　　 B. $\sqrt{\dfrac{e^{-y}}{2}}$ 　　　 C. $\dfrac{e^y}{2}$ 　　　 D. $\dfrac{-e^y}{2}$ 　　　 E. $\sqrt{\dfrac{e^y}{2}}$

Question 309:
Which of the following is largest for $0 < x < 1$

A. $log_8(x)$ 　　 B. $log_{10}(x)$ 　　 C. e^x 　　 D. x^2 　　 E. $\sin(x)$

Question 310:
x is proportional to y cubed, y is proportional to the square root of z. $x \propto y^3, y \propto \sqrt{z}$.
If z doubles, x changes by a factor of:

A. $\sqrt{2}$ 　　 B. 2 　　 C. $2\sqrt{2}$ 　　 D. $\sqrt[3]{4}$ 　　 E. 4

Question 311:
The area between two concentric circles (shaded) is three times that of the inner circle.

What's the size of the gap?
A. r 　　 C. $\sqrt{3}r$ 　　 E. $3r$
B. $\sqrt{2}r$ 　　 D. $2r$ 　　 F. $4r$

Question 312:
Solve $-x^2 \le 3x - 4$
A. $x \ge \dfrac{4}{3}$ 　　 C. $x \le 2$ 　　 E. $-1 \le x \le \dfrac{3}{4}$
B. $1 \le x \le 4$ 　　 D. $x \ge 1$ or $x \ge -4$

Question 313:
The volume of a sphere is numerically equal to its projected area. What is its radius?

A. $\dfrac{1}{2}$ 　　 B. $\dfrac{2}{3}$ 　　 C. $\dfrac{3}{4}$ 　　 D. $\dfrac{4}{3}$ 　　 E. $\dfrac{3}{2}$

Question 314:
What is the range where $x^2 < \dfrac{1}{x}$?
A. $x < 0$ 　　 C. $x > 0$ 　　 E. *None*
B. $0 < x < 1$ 　　 D. $x \ge 1$

Question 315:
Simplify and solve: $(e - a)(e + b)(e - c)(e + d)...(e - z)$?

A. 0 　　　　　　　　　　 D. e^{26} (a+b-c+d...-z)
B. e^{26} 　　　　　　　　 E. e^{26} (abcd...z)
C. e^{26} (a-b+c-d...+z) 　 F. None of the above.

Question 316:
Find the value of k such that the vectors $a = -i + 6j$ and $b = 2i + kj$ are perpendicular.

A. -2

B. $-\frac{1}{3}$

C. $\frac{1}{3}$

D. 2

Question 317:
What is the perpendicular distance between point p with position vector $4i + 5j$ and the line L given by vector equation $r = -3i + j + \lambda(i + 2j)$

A. $2\sqrt{7}$

B. $5\sqrt{2}$

C. $2\sqrt{5}$

D. $7\sqrt{2}$

Question 318:
Find k such that point $\begin{pmatrix} 2 \\ k \\ -7 \end{pmatrix}$ lies within the plane $r = \begin{pmatrix} 2 \\ 3 \\ -1 \end{pmatrix} + \lambda\begin{pmatrix} 4 \\ 1 \\ 0 \end{pmatrix} + \mu\begin{pmatrix} 2 \\ 1 \\ 3 \end{pmatrix}$

A. -2

B. -1

C. 0

D. 1

E. 2

Question 319:
What is the largest solution to $\sin(-2\theta) = 0.5$ for $\frac{\pi}{2} \le x \le 2\pi$?

A. $\frac{5\pi}{3}$

B. $\frac{4\pi}{3}$

C. $\frac{5\pi}{6}$

D. $\frac{7\pi}{6}$

E. $\frac{11\pi}{6}$

Question 320:
$\cos^4(x) - \sin^4(x) \equiv$

A. $\cos(2x)$

B. $2\cos(x)$

C. $\sin(2x)$

D. $\sin(x)\cos(x)$

E. $\tan(x)$

Question 321:
How many real roots does $y = 2x^5 - 3x^4 + x^3 - 4x^2 - 6x + 4$ have?

A. 1

B. 2

C. 3

D. 4

E. 5

Question 322:
What is the sum of 8 terms, $\sum_1^8 u_n$, of an arithmetic progression with $u_1 = 2$ and $d = 3$.

A. 15

B. 82

C. 100

D. 184

E. 282

Question 323:
What is the coefficient of the x^2 term in the binomial expansion of $(2 - x)^5$?

A. -80

B. -48

C. 40

D. 48

E. 80

Question 324:
Given you have already thrown a 6, what is the probability of throwing three consecutive 6s using a fair die?

A. $\frac{1}{216}$

B. $\frac{1}{36}$

C. $\frac{1}{6}$

D. $\frac{1}{2}$

E. 1

Question 325:

Three people, A, B and C play darts. The probability that they hit a bullseye are respectively $\frac{1}{5}, \frac{1}{4}, \frac{1}{3}$. What is the probability that at least two shots hit the bullseye?

A. $\frac{1}{60}$ B. $\frac{1}{30}$ C. $\frac{1}{12}$ D. $\frac{1}{6}$ E. $\frac{3}{20}$

Question 326:

If probability of having blonde hair is 1 in 4, the probability of having brown eyes is 1 in 2 and the probability of having both is 1 in 8, what is the probability of having neither blonde hair nor brown eyes?

A. $\frac{1}{2}$ B. $\frac{3}{4}$ C. $\frac{3}{8}$ D. $\frac{5}{8}$ E. $\frac{7}{8}$

Question 327:

Differentiate and simplify $y = x(x+3)^4$

A. $(x+3)^3$
B. $(x+3)^4$
C. $x(x+3)^3$
D. $(5x+3)(x+3)^3$
E. $5x^3(x+3)$

Question 328:

Evaluate $\int_1^2 \frac{2}{x^2} dx$

A. -1 B. $\frac{1}{3}$ C. 1 D. $\frac{21}{4}$ E. 2

Question 329:

Express $\frac{5i}{1+2i}$ in the form $a + bi$

A. $1 + 2i$ B. $4i$ C. $1 - 2i$ D. $2 + i$ E. $5 - i$

Question 330:

Simplify $7\log_a(2) - 3\log_a(12) + 5\log_a(3)$

A. $log_{2a}(18)$
B. $log_a(18)$
C. $log_a(7)$
D. $9log_a(17)$
E. $-log_a(7)$

Question 331:

What is the equation of the asymptote of the function $y = \frac{2x^2 - x + 3}{x^2 + x - 2}$

A. $x = 0$ B. $x = 2$ C. $y = 0.5$ D. $y = 0$ E. $y = 2$

Question 332:

Find the intersection(s) of the functions $y = e^x - 3$ and $y = 1 - 3e^{-x}$

A. 0 and $\ln(3)$ B. 1 C. $\ln(4)$ and 1 D. $\ln(3)$

Question 333:
Find the radius of the circle $x^2 + y^2 - 6x + 8y - 12 = 0$

A. 3   B. $\sqrt{13}$ C. 5   D. $\sqrt{37}$ E. 12

Question 334:
What value of a minimises $\int_0^a 2\sin(-x)\,dx$?

A. 0.5π B. π C. 2π D. 3π E. 4

Question 335:
When $\frac{2x+3}{(x-2)(x-3)^2}$ is expressed as partial fractions, what is the numerator in the $\frac{A}{(x-2)}$ term:

A. -7   B. -1   C. 3   D. 6   E. 7

END OF SECTION

SECTION 1E: Advanced Physics

Physics Syllabus

- ➤ Calculate, manipulate and resolve Vectors and their components & resultants
- ➤ Calculate the moment of a force
- ➤ Difference between normal and frictional components of contact forces
- ➤ Concept of 'Limiting Equilibrium'
- ➤ Understand how to use the coefficient of Friction including $F = \mu R$ and $F \leq \mu R$
- ➤ Use of the equations of Motion
- ➤ Graphical interpretations of vectors + scalars
- ➤ Derivation + Integration of physical values e.g. Velocity from an acceleration-time graph
- ➤ Principle of conservation of momentum (including coalescence)+ Linear Momentum
- ➤ Principle of conservation of energy and its application to kinetic/gravitational potential energy
- ➤ Application of Newton's laws e.g.
    - ○ Linear Motion of point masses
    - ○ Modelling of objects moving vertically or on a planet
    - ○ Objects connected by a rod or pulleys

Essential Equations

Mechanics & Motion

- ➤ Conservation of momentum $\Delta mv = 0$
- ➤ Force $F = \frac{\Delta mv}{t}$
- ➤ Angular velocity $\omega = \frac{v}{r} = 2\pi f$

Simple Harmonic Motion & Oscillations

- ➤ Displacement $x = A\cos(2\pi ft)$
- ➤ $a = -(2\pi f)^2 x = -(2\pi f)^2 A\cos(2\pi ft)$
- ➤ Speed $v = \pm 2\pi f\sqrt{A^2 - x^2}$

Gravitational forces

- ➤ Force $F = \frac{Gm_1 m_2}{r^2} = \frac{GMm}{r^2}$
- ➤ Potential $V = \frac{Gm}{r}$
- ➤ Acceleration $a = \frac{GM}{r^2} = \frac{\partial V}{\partial r}$

Magnetic fields

- ➤ Magnetic flux $\emptyset = BA$
- ➤ Magnetic flux linkage $\emptyset N = BAN$
- ➤ Magnitude of induced emf $\varepsilon = N\frac{\Delta\emptyset}{\Delta t}$

Radioactivity

- ➤ Decay $N = N_0 e^{-\lambda t}$
- ➤ Half life $T_{1/2} = \frac{ln2}{\lambda}$
- ➤ Activity $A = \lambda N$
- ➤ Energy $E = mc^2$

Current & electricity

- ➤ emf $\varepsilon = \frac{E}{Q} = I(R + r)$
- ➤ Resistivity $\rho = \frac{RA}{l}$
- ➤ Resistors in series $R = \sum_{i=1}^{n} R_i$
- ➤ Resistors in parallel s $R = \sum_{i=1}^{n} \frac{1}{R_i}$

Waves

- ➤ Speed $c = f\lambda$
- ➤ Period $T = \frac{1}{f}$
- ➤ Snell's law $n_1 sin\theta_1 = n_2 sin\theta_2$

Top tip! There is no point doing practice questions until the equations above are second nature to you – most of the advanced physics questions will require multi-step calculations and you simply won't have time to recall equations in the real exam. Thus, it's important to rote learn these so you can be as quick as possible in the exam.

Advanced Physics Questions

Question 336:
A ball is swung in a vertical circle from a string (of negligible mass). What is the minimum speed at the top of the arc for it to continue in a circular path?

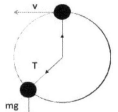

F. 0

G. mgr

H. $2r^2$

I. mg

J. $\sqrt{gr}$

Question 337:
A person pulls on a rope at 60° to the horizontal to exert a force on a mass m as shown. What is the power needed to move the mass up the 30° incline at a constant velocity, v, given a friction force F?

A. $\left(mg + \frac{F}{2}\right)v$

B. $\frac{mg}{\sqrt{2}} - F$

C. $\left(\frac{mg}{2}\right)v$

D. $\sqrt{2}Fv$

E. $\left(\frac{mg}{2} + F\right)v$

Question 338:
What is the maximum speed of a point mass, m, suspended from a string of length l, (a pendulum) if it is released from an angle θ where the string is taught?

A. $2gl(1 - \cos(\theta))$

B. $2gl(1 - \sin(\theta))$

C. $\sqrt{2gl(1 - \cos(\theta))}$

D. $\sqrt{2gl(1 - \sin(\theta))}$

E. $\sqrt{2gl(1 - \cos^2(\theta))}$

Question 339:
Two spheres of equal mass, m, one at rest and one moving at velocity u1 towards the other as shown. After the collision, they move at angles φ and θ from the initial velocity u1 at respective velocities v1 and v2 where $v_2 = 2v_1$. What is the angle θ?

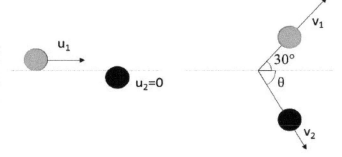

A. 0°

B. sin(50)

C. 30°

D. 45°

E. 90°

Question 340:

The first ball is three times the mass of the others. If this is an elastic collision, how many of the other ball move and at what velocity after the collision?

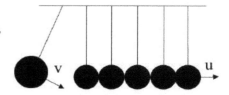

	Number of Balls	Velocity
A	1	3v
B	1	v/3
C	3	v/3
D	3	v
E	3	$\sqrt{v}$

Question 341:

A ball is kicked over a 3 m fence from 6 m away with an initial height of zero. It does not strike the fence and lands 6m behind it. Assuming no air resistance, what is the minimum angle the ball must leave the ground at to make it over the fence?

A. $\arctan(1)$

B. $\arctan\left(\frac{1}{2}\right)$

C. $\arctan(2)$

D. $30°$

E. $\arccos\left(\frac{-1}{2}\right)$

Question 342:

A mass on a spring of spring constant k is in simple harmonic motion at frequency f. If the mass is halved and the spring constant is double, by what factor will the frequency of oscillation change?

A. Stay the same

B. 2

C. 4

D. $\frac{1}{2}$

E. $\sqrt{2}$

Question 343:

A ball is dropped from 3 m above the ground and rebounds to a maximum height of 1 m. How much kinetic energy does it have just before hitting the ground and at the top of its bounce, and what is the maximum speed the ball reaches in any direction?

	E_k at Bottom	E_k at Top	Max Speed
A	0	30m	$2\sqrt{15}$
B	0	30m	30
C	30m	0	$2\sqrt{15}$
D	30m	0	60
E	60m	0	60

Question 344:

An object approaches a stationary observer at a 10% of the speed of light, c. If the observer is 2 m tall, how tall will it look to the object?

A. $0.9l$ B. $1.1l$ C. l D. $l\sqrt{0.99}$ E. $l\sqrt{1.01}$

Question 345:

What is the stopping distance of a car moving at v m/s if its breaking force is half its weight?

A. v^2 B. $\frac{v^2}{g}$ C. $2mv$ D. $\frac{v^2}{2}$ E. $\sqrt{mg}$

Question 346:
The amplitude of a wave is damped from an initial amplitude of 200 to 25 over 12 seconds. How many seconds did it take to reach half its original amplitude?

A. 1   B. 2   C. 3   D. 4   E. 6

Question 347:
Two frequencies, f and $\frac{7}{8}f$, interfere to produce beats of 10 Hz. What is the original frequency f?

A. 11 Hz   B. 60 Hz   C. 80 Hz   D. 160 Hz   E. 200 Hz

Question 348:
Refer to the diagram shown to the right. The two waves represent:

A. A standing wave with both ends fixed
B. The 4th harmonic
C. Destructive interference
D. A reflection from a plane surface
E. All of the above

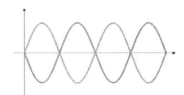

Question 349:
Radioactive element $_b^a X$ undergoes beta decay, and the product of this decay emits an alpha particle to become $_d^c Y$. What are the atomic number and atomic mass?

	c	d
A	a-4	b+1
B	a-3	b-2
C	a-4	b-1
D	a-5	b
E	a-1	b-4

Question 350:
A gas is heated to twice its temperature (in Kelvin) and allowed to increase in volume by 10%. What is the change in pressure?

A. 82% increase   C. 110% increase   E. 40% decrease
B. 90% increase   D. 18% decrease   F. 82% decrease

Question 351:
A beam of alpha particles enters perpendicular to the magnetic field, B, shown below coming out of the page and is deflected to follow path T. What path would a beam of electrons follow?

A. P
B. Q
C. R
D. S

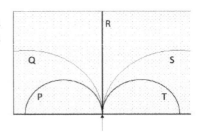

Question 352:
If all light bulbs and power supplies are identical and the joining wires have negligible resistance, which of the following light bulbs will shine brightest?

A. 1,2,3
B. 6
C. 7,8,9
D. 1,2,3,6
E. 4,5

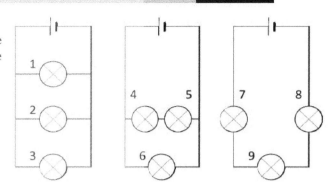

Question 353:
A convex lens with focal length f is used to create an image of object O. Where is the image formed?

A. V
B. W
C. X
D. Y
E. Z

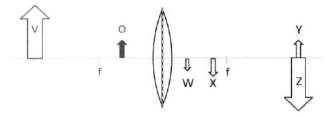

Question 354:
A flower pot hangs on the end of a rod protruding at right angles from a wall, held up by string attached two thirds of the way along. What must the tension in the string be if the rod is weightless and the system is at equilibrium?

A. $mg \sin \theta$

B. $\dfrac{3mg}{2\sin \theta}$

C. $\dfrac{3mg}{2\cos \theta}$

D. $\dfrac{2mg}{3\sin \theta}$

E. $\dfrac{2mg}{3\cos \theta}$

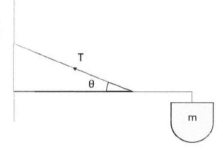

Question 355:
When a clean, negatively charged metal surface is irradiated with electromagnetic radiation of sufficient frequency, electrons are emitted. This observation describes what phenomenon and what property does the intercept on the experimental plot below represent?

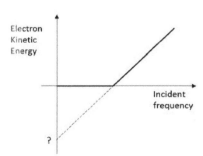

A		
B	Evaporation	Stopping potential
C	Wave particle duality	Work function
D	Thermionic emission	Work function
E	Photoabsorbtion	Stopping potential

Question 356:
Which of these wave phenomena can be explained by Huygen's principle of wave propagation?

1. Diffraction
2. Refraction
3. Reflection
4. Interference
5. Damping

A. None   B. 1,2,3   C. 1,4   D. 1,2,3,4   E. All

Question 357:
What is the maximum efficiency of an engine where the isothermal expansion of a gas takes place at $T_1 = 420$ K and the reversible isothermal compression of the gas occurs at temperature $T_2 = 280$ K?

A. 43%   B. 57%   C. 75%   D. 92%   E. 100%

Question 358:
Which of the following statements is true?

A. A capacitor works based on the principle of electromagnetic induction.
B. A motor requires a AC input
C. Transformers produce DC current
D. The magnetic field produced by a current carrying wire is parallel to the direction of flow of charge.
E. A generator must have a moving wire.

Question 359:
For the logic gate below, if inputs are all set to 1, what would the value of X, Y and Z be?

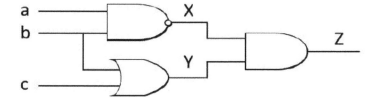

	X	Y	Z
A	0	0	0
B	0	0	1
C	0	1	0
D	1	0	0
E	1	1	1

Question 360:
What material property is given by each point P, Q and R on a stress-strain curve?

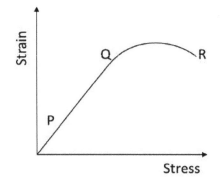

	P	Q	R
A	Elastic Modulus	Yield stress	Fracture toughness
B	Tensile Modulus	Plastic onset	Yield stress
C	Hardness	Stiffness	Ductile failure
D	Ductility	Elastic limit	Brittle fracture
E	Young's Modulus	Yield stress	Fracture stress

SECTION 2

Section 2 questions are designed to stretch you by putting you out of your comfort zone. In addition to the core knowledge required for section 1, you will need to apply core scientific principles in unfamiliar contexts.

All section 2 questions require you to have the knowledge from section 1A maths as well as the corresponding subject knowledge from section 1. For example biology section 2 questions require 1A maths and 1D Biology. Similarly, physics Section 2 questions require 1A maths, 1B physics and 1E advanced maths/physics. Each question is out of 25 marks.

Specialise

The very nature of the exam allows you to tailor your answers to your preferences and your abilities with regards to the different topic areas. This applies for both sections of the exam. You are expected to answer the maths questions (part 1), but other than that, you can choose freely which parts you want to answer. In section 2 you are even freer being able to choose any 2 questions of 6 covering Biology, chemistry and Physics. With regards to preparation, this means you should prepare to be able to apply your math skills, but the other parts should be easier to prepare for. It may for this reason pay of to "specialise" on specific topic areas. In general, it is best to focus on the same topics for both sections.

Thrive on Adversity

The NSAA is specifically designed to be challenging and to take you out of your comfort zone. This is done in order to separate different tiers of students depending on their academic ability. The reason for this type of exam is that Cambridge attracts excellent students that will almost invariable score well in exams. If, for this reason, during your preparation, you come across questions that you find very difficult, use this as motivation to try and further your knowledge beyond the simple school syllabus. This is where the option for specialisation ties in.

Be Efficient

You have 40 minutes to answer 2 questions that may be text based and include graphs. This leaves you with roughly 20 minutes per question. It is vital that you manage your time well. The best way to deal with this is to shorten the time you give yourself during your preparation progressively until you are able to finish your Section 2 question within roughly 30 minutes. This will give you the confidence for the exam. Another way to improve your performance in section 2 is to get familiar with different types of graphs and how to represent scientific information efficiently. This will help you to be as efficient as possible.

Practice Calculus

Section 2 does allow you to use calculators. For this reason, practicing calculus might seem unnecessary. But it will give you a big advantage to be confident with it as you will be significantly more efficient and faster. As the math part is core knowledge for all section 2 questions, you can be certain that to some degree or other you will be required to apply your calculus. The better you prepare for your section 1 maths part, the better you will fare in section 2.

Don't be afraid to show off!

Examiners in the NSAA are not only looking for someone who has the correct answers to the questions being asked, but also for someone who has that extra spark. There may be some open-ended questions in the NSAA, which leave room for the student to input any extra knowledge on the subject they may have. Do not be afraid to use any extra knowledge that you have!

Section 2: Physics

Think in Applied Formulas

Trying to answer section 2 physics questions without first learning all the core formulas is like trying to run before you can walk – ensure you're completely confident with all the formulas given in sections 1B + 1E before starting the practice questions.

Graphs

Graph sketching is usually a tricky area for many students. When tackling a graph sketching problem, there are many approaches, however, it is useful to start with the basics:

➢ What is the value of y when x is zero?
➢ What is the value of x when y is zero?
➢ Are there any special values of x and y?
➢ If there is a fraction involved, at what values of x is the numerator or denominator equal to zero?

If you asked to draw a function that is the sum, product or division of two (sub) functions, start by drawing out all of the sub functions. Which function is the dominant function when x > 0 and x < 0.

Answering these basic questions will tell you where the asymptotes and intercepts etc. are, which will help with drawing the function.

Remember the Basics

A surprisingly large number of students do not know what the properties of basic shapes etc. are. For example, the area of a circle is πr^2, the surface area of a sphere is $4\pi r^2$ and the volume of a sphere is $\frac{4}{3}\pi r^3$. To 'go up' a dimension (i.e. to go from an area to a volume) you need to integrate and to 'go down' a dimension you need to differentiate. Learning this will make it easy to remember formulas for the areas and volumes of basic shapes. It is also important to remember important formulas even though they may be included in formula booklets. The reason for this is that although you may have access to formula booklets during exams, this will not be the case in interviews which will follow the NSAA. In addition, flicking through formula booklets takes up time during an exam and can be avoided if you are able to memorise important formulas.

Variety

Physics is a very varied subject, and the questions you may be asked in an exam will reflect this variety. Many students have a preferred area within physics, e.g. electronics, astrophysics or mechanics. However, it is important to remember not to neglect any subject area in its entirety on the basis that you will answer questions on another subject. It is entirely possible that of the two physics questions in the NSAA, both will be on subjects outside of you comfort zone. As it is common for students who study physics not to study biology, this would be a bad situation as you would have a very limited number of questions left to choose from.

Show Working

It is very important to show all working clearly, and not just write the final answer. If you compare a physics question to an English or history essay, most people would agree that an essay would not simply consist of the writer's final opinion and conclusions. For the same reasons, when answering any physics problem, you must show all working clearly so that examiner can see how you arrived at an answer, even if the answer is incorrect. When calculating an answer, it is strongly advisable to write out everything algebraically as this is the clearest way in which working can be written.

For example, the examiner will not know what $\frac{(6.67\times10^{-11})\times3.14^2\times28.6581}{12.9/0.0036}$ means, but they will understand what $\frac{G\pi^2\rho}{\sigma/\alpha}$ means (if the numbers correspond to the symbols). Furthermore, if you are to make a silly mistake (e.g. you mistype something into you calculator at the end etc.), the examiner will be able to give you credit for writing the correct method, which may amount to most of the marks available in a question even if your final answer is incorrect. This is not possible if you write everything out numerically, as the examiner will not have the time or inclination to spend a long time over deciphering your working to see where you went wrong.

Similarly, it's very important that you clearly state any assumptions that you're making e.g. when calculating the gravitational force between two masses do you assume the masses are point masses?

Physics Question 1

The Earth receives radiation from the Sun which is both absorbed and reflected from the Earth's surface and atmosphere. The solar flux has a value of Ω.

a) Draw a diagram showing the distribution of solar flux across the Earth's surface. Use this to calculate the proportion of solar flux incident on the Earth's surface.

[7 marks]

b) The solar flux incident on the Earth's surface is absorbed and converted to infrared radiation which is emitted away from the surface. Objects that are maximally efficient absorbers are called blackbodies. The relationship between the temperature of blackbodies and their rate of radiation is given by the Stefan-Boltzmann law,

$$F = \sigma T_0^4$$

Where F is the total emitted thermal radiation, σ = Stefan-Boltzmann constant = 5.67×10^{-8} $Jm^{-2}K^{-4}$ and T_0 is the absolute temperature of the blackbody.

Solar energy is also reflected by the atmosphere. The ratio of solar radiation reflected by the atmosphere to total incoming solar radiation is called the albedo, a.

Write expressions for the total radiation that is reflected by the Earth's atmosphere and the radiation emitted as infrared radiation from the Earth's surface.

[2 marks]

c) Using you answers to parts a and b, calculate the surface temperature of the Earth in terms of the solar flux, Ω.

[3 marks]

d) Solar flux, Ω, of the Sun's unimpeded rays at a distance equal to the distance between Sun and the Earth is 1372 Wm^{-2}, and the albedo of the Earth is 0.3. Use these values to calculate the blackbody temperature of the Earth.

[1 mark]

e) Is the value of T_0 greater or lower than what you would expect? What factors have not been considered during your calculations?

[5 marks]

f) This distance between the Sun and the Earth is 1.5×10^{11} m and the radius of the Sun is 6.96×10^8 m. Use these values and the value of Ω given above to calculate the blackbody temperature of the Sun.

[5 marks]

g) What assumptions have you made in the calculation above?

[2 marks]

[Total: 25 Marks]

Physics Question 2

a) Sketch the graph $y = \dfrac{\sqrt{x^2+1}-\sqrt{x^2-1}}{\sqrt{x^4-1}}$

[8 marks]

b) Evaluate the integral $\int \dfrac{\sqrt{x^2+1}-\sqrt{x^2-1}}{\sqrt{x^4-1}}\,dx$

[10 marks]

c) If the velocity of a train, $v(t)$ is given by:

$$v(t) = \begin{cases} 5t & 0 < t < 1.5 \\ \dfrac{\sqrt{t^2+1}-\sqrt{t^2-1}}{\sqrt{t^4-1}} & 1.5 < t < 2.0 \\ 2t^2 - 5t & 2.0 < t < 2.5 \end{cases}$$

Where t = time in hours and v = velocity in km s^{-1}.
Sketch a graph showing velocity as a function of time, t.

[4 marks]

d) Calculate the distance travelled by the train between $t = 0$ and $t = 2.5$ hours.

[3 marks]

[Total: 25 Marks]

Physics Question 3

Radioactive decay of any radiogenic atom is a random process that is independent of neighbouring atoms, physical conditions and the temperature state of the atom. The probability of a radiogenic isotope undergoing radioactive decay is called the decay constant, λ, and is different for every isotope.

a) If at time T there are t atoms of a radioactive isotope, and at time $T + \partial t$ there are ∂N atoms have decayed, derive an equation that relates the number of atoms N at time t to the decay constant, λ.

[5 marks]

b) The rate of decay is also called the activity, A. Use your answer from part a to express the activity as a function of time, t.

[1 mark]

c) At time $t = t_{1/2}$, half of the original atoms of a radioactive isotope will be present and the half will have undergone radioactive decay. At this time, $A = \frac{1}{2}A_0$, and time $t_{1/2}$ is known as the radioactive half-life. Use your answer to part a to express $t_{1/2}$ as a function of λ.

[3 marks]

d) (i) Complete the empty column in the table below:

[4 marks]

Parent Isotope	Daughter Isotope	Decay Products	λ (a^{-1})
^{238}U	^{206}Pb		1.55×10^{-10}
^{235}U	^{207}Pb		9.85×10^{-10}
^{232}Th	^{208}Pb		4.95×10^{-11}
^{87}Rb	^{87}Sr		1.42×10^{-11}
^{147}Sm	^{143}Nd		6.54×10^{-12}
^{40}K	^{40}Ca & ^{40}Ar		4.95×10^{-10} & 5.81×10^{-11}
^{39}Ar	^{39}Ar		2.57×10^{-3}
^{176}Lu	^{176}Hf		1.94×10^{-11}
^{187}Re	^{187}Os		1.52×10^{-11}
^{14}C	^{15}N		1.21×10^{-4}

(ii) If the total number of daughter isotope atoms present at time t is D, write an expression that expresses t as a function of D and N.

[3 marks]

(iii) State and explain which of the parent isotopes listed in the table would be best suited to date:

→ Age of the Earth **[2 marks]**
→ Ancient artefacts **[2 marks]**

(e) An igneous rock contains 11.7×10^{-5}g of ^{238}U and 3.58×10^{-5}g of ^{206}Pb and a negligible amount of ^{208}Pb. Stating your assumptions, determine the approximate age at which the rock formed.

[5 marks]
[Total: 25 Marks]

Physics Question 4

a) For the two point masses m_1 and m_2 at a distance r apart, the gravitational force of attraction between the two masses is given by: $F = \frac{G m_1 m_2}{r^2}$

Whereby $G = 6.67 \times 10^{-11} \, \text{m}^3 \, \text{kg}^{-1} \, \text{s}^{-1}$. The gravitational potential, V due to mass m_1 is defined as $V = -\frac{G m_1}{r}$

(i) What is the gravitational potential energy of a mass m_2 at distance r from mass m_1?

[1 mark]

(ii) What is the relationship between gravitational potential, V, and gravitational acceleration, a, of mass m_2 towards m_1?

[2 marks]

b) (i) For a spherical shell of radius b, calculate the area of a circular strip (hint: if θ is the angle between the centre of the shell and the surface, and $d\theta$ is the difference in angle between two nearby points on the outside of the shell, then $d\theta \ll \theta$)

[3 marks]

(ii) Assume the shell has a thickness t and uniform density ρ. Given that for a distribution of masses, the gravitational potential is given by: $V = -G \int_m \frac{dm}{r}$

Calculate the gravitational potential due to the shell at an arbitrary point P at distance D away from the shell, for cases when P is inside and outside of the shell.

[15 Marks]

(iii) Is the gravitational potential of position P for a point inside the shell dependent or independent of position?

[1 mark]

(iv) What is the gravitational acceleration of a point P inside the spherical shell?

[1 Mark]

(v) What is the gravitational acceleration at a point which is at a distance r from the centre of a sphere of radius b (where $\ll r$) and of density ρ?

[2 marks]

[Total: 25 Marks]

Physics Question 5

a) (i) What is meant by the term 'electromagnetic induction'?

[2 marks]

(ii) Using diagrams, state and explain Lenz's law in relation to the electromagnetic induction

[5 marks]

(iii) State Faraday's law, and write down equations that show its relation to Lenz's law and the voltage of the generated current

[5 marks]

(iv) Calculate the voltage generated when a rectangular coil of length 5 cm, width 10 cm and 35 turns is rotated at a rate of 50 rotations per minute through a magnetic field of strength 0.4T

[3 marks]

b) (i) Describe the dynamo theory as an explanation to the presence of the Earth's magnetic field

[3 marks]

(ii) Planets such as Mars have very weak magnetic fields. What would the effect of losing the Earth's magnetic field be on the planet? What would this mean with regards to the internal structure of the Earth

[5 marks]

(iii) Many planetary bodies which are not thought to have an internally-generated magnetic dynamo still have magnetic fields. Give an example of such a planetary body and explain how its dynamo is generated.

[2 marks]

[Total: 25 Marks]

Section 2: Chemistry

Keep things Simple

Chemistry is a science of simplification. The more flexible you are with the application of your knowledge, the better you will be able to apply it to the changing environment of the questions. This will allow you to use principles and apply them for various questions helping with time conservation.

Remember Applications

Chemistry is dominated by applications in the real world and you can expect the questions to reflect this. If you design your revision round this idea, it will help you not only prepare better as you are more likely to retain the information, but it will also make sure that you are not thrown off by challenging application questions.

Analyse the Questions

There are two main types of chemistry questions: organic and inorganic. Organic chemistry questions often involve drawing and naming molecules, whilst inorganic chemistry questions can be more mathematical. It is paramount that you analyse each question in advance to ensure that you are aware of what is coming up. You do not want to be halfway through a question and realise that you would rather have answered another question!

Focus on Clarity

Chemistry often involves drawing things like molecules and writing out sometimes long chemical reactions. It is very important that your writing is clear and legible - a wayward dash or heavy smudge may be taken by an examiner to be a minus sign etc. It is worth buying a mechanical pencil and rubber ahead of an exam and having spare leads.

Memorise Formulas

It goes without mentioning the importance of basic formulas, such as the gas law ($pV = nRT$), molar amount ($n = M/M_r$) etc. Although these formulas are provided in formula books, many students waste time during an exam looking these up. Such basic equations should become second-nature to you as your revision progresses, and with practice you should be able to apply the correct formulas to problems automatically.

Practice makes Perfect

Perhaps the one of the trickiest questions possible in a chemistry exam is being asked to draw a complex molecule. There is an almost unlimited amount of material on the internet which can be used to prepare from. What can also be helpful to do is to slightly alter examples from books/the internet (e.g. move a methyl group to another carbon atom etc.) and re-draw a molecule. This way, you ensure that you fully understand how nomenclature in chemistry works. It is also a good idea to memorise other standard chemical processes/reactions, such as the free radical breakdown of ozone etc. which is available freely online.

Chemistry Question 1

Compounds added to pure water react and undergo changes in concentration until the products reach equilibrium. If two reactants, **A** and **B** are added to pure water, the reaction that takes place is: $A + B \leftrightarrow C + D$

a) When the products and reactants are at chemical equilibrium, describe the rates of the forward and backward reactions and concentration of the reactants and products.

[2 marks]

b) Write three balanced equations describing the reactions that occur when CO_2 is dissolved in water.

[3 marks]

c) When the reaction between the reactants and products of the above reaction has reach chemical equilibrium, the following relationship describes the concentration of the products and reactants: $\frac{[C][D]}{[A][B]} = 10^{-pK}$

Whereby pK is the dissociation constant that characterises this relationship for a specific reaction and is determined through experimental research. The table below lists the values for the Henry's constant, pK, for the reactants and products of the reaction of part b:

Reaction	10^{-pK}
$H_2CO_3 \leftrightarrow H^+ + HCO_3^-$	$10^{-6.35}$
$HCO_3^- \leftrightarrow H^+ + CO_3^{2-}$	$10^{-10.33}$
$H^+ + OH^- \leftrightarrow H_2O$	10^{-14}

At the equator, atmospheric CO_2 has a partial pressure, $p[CO_2]$, of 340 ppm(v). The concentration of atmospheric CO_2 is given by the following relationship: $\frac{[H_2CO_3]}{p[CO_2]} = 10^{1.47}$ molecules/litre
Use the values from the table to calculate the concentration of H_2CO_3.

[1 mark]

d) In order to calculate the pH of pristine precipitation, use the relationship: $[H^+] = [HCO_3^-] + 2[CO_3^{2-}] + [OH^-]$ to write down equations that express $[H^+]$ in terms of the products of the above reaction.

[8 marks]

e) For $[H^+]^n = 10^X$, $[H^+] = 10^{X/n}$. Using an appropriate approximation, calculate the concentration of H^+ of CO_2 in equilibrium with the Earth's oceans.

[2 marks]

f) What is the relationship between pH and the concentration of H^+?

[1 mark]

g) Use your answers to calculate the pH of rain water in equilibrium with atmospheric CO_2.

[1 mark]

h) CO_2 is not the only gas that contributes to the acidification of precipitation. Sulphur dioxide, both naturally and anthropogenically produced also contributes to acid rain.

Reaction	10^{-pk}
$H_2SO_3 \leftrightarrow H^+ + HSO_3^-$	$10^{-1.77}$
$HSO_3^- \leftrightarrow H^+ + SO_3^{2-}$	$10^{-7.21}$

The concentration of atmospheric SO_2 is 0.2 ppb(v), and $\frac{[H_2SO_3]}{p[SO_2]} = 10^{0.096}$ molecules/litre. The charge-balance equation for the combined $CO_2 + SO_2$ systems is: $[H^+] = [HCO_3^-] + 2[CO_3^{2-}] + [HSO_3^-] + 2[SO_3^{2-}] + [OH^-]$

Using the values above and the approximation equation from part e, calculate a new value for the pH of rainwater that is in equilibrium with the atmosphere.

[5 marks]

i) What are the main sources of atmospheric SO_2?

[2 marks]

[Total: 25 marks]

Chemistry Question 2

a) Draw shapes representing the distribution of electron density in p-orbitals around atom X.

[3 marks]

b) (i) Species XH_4 can carry a zero, a positive or a negative charge. Write formulas for **reactions between XH_3 and H ions,** and draw dot and cross diagrams representing the products, stating their charge.

[6 marks]

(ii) What shape is the molecule XH_4, and what are the bond angles?

[2 marks]

(iii) State which atoms X can be

[3 marks]

c) An aqueous solution of 2,2 di-methyl butanoic acid with a concentration of 0.05 mol dm^{-1} has a pH of 3.2.

(i) Draw the structure of the acid and write its chemical formula

[3 marks]

(ii) Calculate the pK_a of the acid, showing your working

[3 marks]

(iii) Phenol has a pK_a of 7.5. Draw the molecule phenol (C_6H_5OH) and calculate the pH of a solution containing phenol and 2,2 di-methyl butanoic acid in equal concentrations of 0.1M.

[5 marks]

[Total: 25 Marks]

Chemistry Question 3

a) Draw full diagrams showing every atom and bond for the following molecules:

(i) 3-dimethyl butane
(ii) Heptanal
(iii) Methyl propanoate
(iv) Ethanenitrile
(v) 2-bromo,3-chloro butane

[10 marks]

b) (i) What process is used in the oil industry to break down long-chain hydrocarbon molecules into shorter-chain hydrocarbon molecules?

[1 mark]

(ii) Write down an equation and draw the structures of the broken-down products of decane, $C_{10}H_{22}$.

[3 marks]

(iii) In what order will the following products be produced during the breakdown of long-chain hydrocarbons?

Gasoline, diesel oil, bitumen, lubricants, petroleum gases and kerosene

[2 marks]

(iv) In industry, catalysts are used to speed up the thermal decomposition of long-chain hydrocarbons. What are these catalysts called?

[1 mark]

(v) In the presence of chlorine, ethane may undergo a free radical reaction to produce chloroethane. State what reaction conditions are necessary, and write out and name each step of this reaction mechanism.

[6 marks]

(vi) What impurities are commonly found in naturally produced hydrocarbons, and what are the main risks involved in the process of breaking down long-chain to short-chain hydrocarbons?

[2 marks]

[Total: 25 Marks]

Chemistry Question 4

a) (i) What is isomerism?

[2 marks]

(ii) Draw all the isomers of the molecule C_4H_8

[12 marks]

(iii) Which of the molecules you have drawn above are structural isomers and stereoisomers and why?

[2 marks]

b) Write out the mechanisms for the following reactions:
(i) Ethene and HBr
(ii) Cyclohexene and HBr

[8 marks]

(iii) What is this process called?

[1 mark]

[Total: 25 Marks]

Chemistry Question 5

a) Organic compounds can react with the compound H–X (where X = halogen), however the rate of reaction varies. Why does the rate of reaction vary depending on X and how does this relate to bond strength? Write the different H–X molecules in order of reactivity, with the most reactive molecule first.

[4 marks]

b) Elimination reactions are a type of organic reaction in which two substituents are removed from a molecule in either a one step or two step reaction.

(i) Write the one-step reaction mechanism between isobutyl bromide ($(CH_3)_3CBr$) with potassium ethanoxide ($CH_3CH_2O^-K^+$), dissolved in ethanol.

[5 marks]

(ii) Write out the two-step reaction between isobutyl bromide and potassium ethanoxide, dissolved in ethanol. Label clearly all hydrogen bonds present in the transitional stage.

[6 Marks]

c) (i) Write a relationship between pK_a and ethanoic acid, CH_3COOH

[2 marks]

(ii) Calculate the pH of 0.017 M of aqueous ethanoic acid, where $pK_a = 4.76$.

[4 marks]

(iii) Explain why the pK_a of ethanoic acid is different to the pK_a of ethanol, where $pK_a = 16.0$ for ethanol.

[4 marks]

[Total: 25 Marks]

Section 2: Biology

Preparation

Biology questions are heavily knowledge based and require you to apply a principle in order to get to the correct answer. Thus, it essential you're familiar with concepts like genetics, natural selection, cells, organs systems and species interaction. It will make your life a lot easier if you feel comfortable with all the different aspects of biology. Specialisation here can go a long way. Chances are that if you prepare well for as many of the different topics as possible you will not be caught off guard and you will not have to waste time on trying to recall them.

Think in Principles

Tying in with the previous point, thinking in principles in biology will help you select relevant information and separate it from irrelevant information. This will allow you to work efficiently and you will be able to produce high quality answers. Try and figure out why a certain system makes sense from an efficiency perspective.

Remember, biology is about producing the maximum result with a minimum of energy expenditure. It can sometimes also help to use case studies for revision. This will allow you to study an entire variety of biological topics under one common headline that will make it easier to think through the context. For example, diabetes lets you revise hormones associated with digestion and energy control as well response mechanisms to cellular stimulation and metabolism.

Language

Get used to using scientific language. The more precise you can be with your answers, the less time it will take you to convey your information meaning you will be that much more likely to answer all the questions. Technical terms are a great tool as they are so precise and make you look a lot more professional. In this context however you MUST ensure that you are using the technical terms correctly! Using technical terms incorrectly is a very quick way to getting the question completely wrong, even if you might be in the right path.

Logic is Key

The very nature of the exam is designed to throw you off guard and force you to use your knowledge to come to conclusions you might not have been taught at school. This is what separates good and great students. Applying the principles you have learned, securely and efficiently, will allow you to answer most questions correctly, even if you have never addressed them before.

Biology is a very logical science that is largely based on the idea that the goal is always to reach the maximum amount of effect with a minimum of energy expenditure. Thus, it is obvious that you need to have a sound scientific basis in order to recognise the different connections between topics.

Time = Marks

Due to the large amount of information that can be examined in the biology questions, it is essential to keep an eye on your time. You have about 20 minutes per question which means that it is easy to run out of time, especially if you are required to make a drawing or if there are large amounts of text resource to read. One catch to be aware of is that there can be some degree of interconnection between different sub-questions making it necessary to think through one first part before progressing to the next.

Biology Question 1:

In healthy humans, the blood sugar levels are controlled by the secretion of insulin from the pancreas. During episodes of high blood sugar levels, the beta cells in the pancreas secrete insulin, a protein hormone, which allows most cells to take up sugar from the blood.

A. The following DNA sequence is part of the human insulin gene:

3' ... GAC ACG CCG AGT ... 5'
5' ... CTG TGC GGC TCA ... 3'

Using the decoding wheel, what amino acid sequence does this encode?

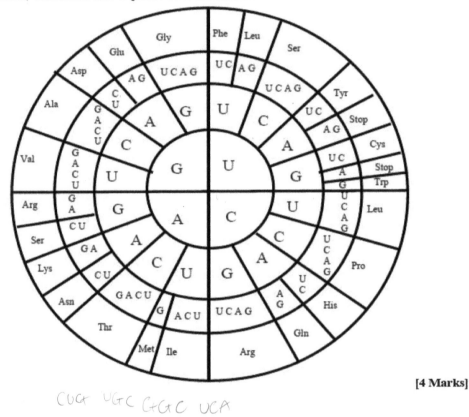

[4 Marks]

CUG UGC GGC UCA

B. Would a point mutation on the first space of the first base triplet be possible without changing the original amino acid sequence?

[2 Marks]

C. In the past, insulin was derived from slaughtered animals such as pigs and cows. Introducing this foreign protein to the human body resulted in a variety of hypersensitivity reactions. After isolating the insulin gene in the human genome, mass production in bacteria was possible.

How would you produce human insulin in bacteria?

[6 Marks]

D. Hyperinsulinemia is a disorder commonly associated with individuals using insulin for therapeutic means in the context of diabetes mellitus. It can also occur in the context of a genetic disease affecting the natural control mechanisms of insulin release.

Understanding that insulin release is controlled through the binding of glucose to a beta cell surface receptor which causes the release of insulin by triggering vesicle release through an amplification protein, where would you expect the mutation to be located?

[4 Marks]

E. Insulin acts by allowing sugar to enter cells. Explain what happens in type 2 diabetes using the drawing below (drawing displayed to inform about GLUT-4 translocation in response to insulin receptor signalling):

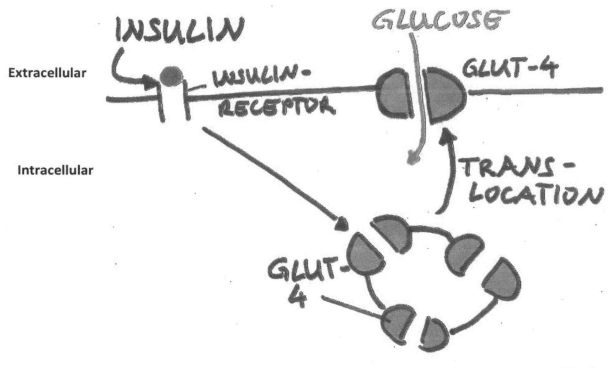

[6 Marks]

F. Glucagon is another hormone produced by the pancreas. What action does glucagon have on liver cells?
[3 Marks]

[Total: 25 Marks]

Biology Question 2:

Many insects possess a stinger for personal defence against predators. This includes insects such as bees and wasps. In nature, it often happens that frogs, especially if not confronted with stinging insects in the past, attempt to eat these insects.

The hunting behaviour of frogs involves them sitting motionless, waiting for their prey to move past them. Hunting happens through rapid extension of the tongue, grabbing of the prey and swallowing in one fluent and extremely rapid motion. Attempting to eat an insect armed with a stinger leads to the frog immediately spitting out the prey followed by retching and attempts to remove the insect from the mouth, despite it being gone already.

A. What differences are there between the nerve fibres of insects and vertebrates, such as frogs, and how does this affect the speed at which information is transmitted?

[4 Marks]

B. Draw a schematic representation of a reflex arc.

[4 Marks]

C. The perception of pain is due to pain receptors in the skin. The perception of pain is relayed to the spinal cord as well as to the brain. In the spinal cord, the pain impulse produces a specific reflex. Describe the purpose of relaying the pain impulse both to the spine and the brain.

[4 Marks]

D. Describe the evolutionary basis for mimicry amongst poisonous and non-poisonous species.

[3 Marks]

E. Pain impulses is transmitted along the nerve fibre through action potentials which are maintained by ion channels. Describe how action potentials are generated.

[6 Marks]

F. Lidocaine is a local anaesthetic used to control pain in many medical environments. It acts by blocking sodium channels. Explain how lidocaine blocks transmission of pain signals along nerve fibres.

[4 Marks]

[Total: 25 Marks]

Biology Question 3:

Chimpanzees live in large, hierarchical groups of up to 100 individuals. They are omnivores and eat plant materials as well as insects. They also hunt for small mammals. The large groups of chimps, often split up into smaller groups for food gathering and hunting and it has been observed that the hunting success rate increases proportionally with the size of the subgroups - in groups, larger than 6 individuals, success rate approaches roughly 90%.

A. Describe how large units like this group of chimpanzees are organized.

[4 Marks]

B. What are the evolutionary advantages of organisation of individuals in large groups?

[4 Marks]

C. Many animals living in groups have developed forms of communication. What is the purpose of being able to communicate and do you think there is any limitation on the type of species developing communication mechanisms?

[4 Marks]

D. In some animals, such as certain mountain goat species, reproduction is limited to a few weeks each year. This increases reproductive pressure on both males and females in the population. What reasons can you see for this behaviour?

[3 Marks]

E. What role does genetic diversity play in the survival of populations?

[3 Marks]

F. Provide a hypothesis why most animal groups are organized in a male dominance pattern and female dominance is rarer.

[7 Marks]

[Total: 25 Marks]

Biology Question 4:

Palicourea marcgravii is a plant occurring in South America. Ingestion of this plant causes sudden death in ruminants such as cows but not in monogastriers such as horses where it causes a more slowly progressing disease. The symptoms of the sudden death variant include muscle cramps, shortness of breath and abnormal movements of extremities followed by death within roughly 10 minutes.

A.   Formulate a hypothesis why the sudden death variant of the reaction is exclusive to ruminants.

[4 Marks]

B.   What action does Adrenaline have on organs such as the heart, the gut, the lungs and the blood vessels?

[3 Marks]

C.   Recently the toxic ingredients of the plant have been isolated. One of them, fluoroacetate, directly interferes with the production of ATP in the citric acid cycle. The third chemical component, N-Methyltyramine, is not by itself toxic but accelerates the toxic action of fluoroacetate by acting as a competitive substrate to Monoamine oxidase A (MAO-A) and thereby increasing adrenalin concentration in the body. Explain how this causes death.

[5 Marks]

D.   How does enzyme inhibition via competitive antagonist work?

[3 Marks]

E.   Suggest a possible treatment ingestion of Palicourea marcgravii.

[3 Marks]

F.   Describe the purpose of the fight or flight with regards to the physiologic changes associated with it.

[7 Marks]

[Total: 25 Marks]

Biology Question 5:

In *E. coli*, the expression of different enzymes can be controlled by operons that rapid to the presence of specific substrates in the environment. One example of this, is the response to arabinose. In absence of arabinose, the digesting enzymes are not being synthesized as the RNA polymerase cannot bind to a promotor triggering expression of the required gene. In the presence of arabinose in the environment, it binds to a regulator protein which in turn binds to the promotor causing activation of the enzyme gene.

A. Draw a schematic representation of the expression control in presence of arabinose.

[3 Marks]

B. Using the information provided above, formulate a hypothesis of how E coli controls the expression of lactase.

[2 Marks]

C. How would a gain of function mutation in the operon affect enzyme expression and what consequences would this have for the bacterium?

[3 Marks]

D. Why do bacteria have mechanisms to control enzyme production based on environmental substrates?

[5 Marks]

E. Operon mediated enzyme expression also plays a role in facultative antibiotic resistance in certain bacteria. What do you think this means?

[5 Marks]

F. Why is operon-controlled expression mechanisms less common in humans?

[7 Marks]

[Total: 25 Marks]

ANSWERS

Answer Key

Question	Answer	Question	Answer	Question	Answer	Question	Answer
1	B	51	D	101	D	151	D
2	C	52	D	102	G	152	F
3	C	53	B	103	F	153	A
4	C	54	E	104	B	154	A
5	E	55	E	105	A	155	D
6	A	56	B	106	C	156	C
7	C	57	C	107	G	157	B
8	E	58	A	108	D	158	F
9	E	59	C	109	D	159	A
10	C	60	B	110	E	160	C
11	E	61	B	111	B	161	C
12	E	62	B	112	A	162	E
13	E	63	C	113	E	163	F
14	B	64	C	114	D	164	G
15	C	65	A	115	F	165	F
16	B	66	C	116	E	166	A
17	B	67	D	117	A	167	C
18	C	68	C	118	B	168	B
19	D	69	D	119	C	169	D
20	C	70	A	120	D	170	B
21	B	71	C	121	F	171	E
22	A	72	B	122	E	172	B
23	F	73	B	123	C	173	D
24	D	74	A	124	C	174	D
25	A	75	C	125	B	175	D
26	B	76	F	126	D	176	D
27	A	77	A	127	F	177	E
28	F	78	D	128	G	178	E
29	D	79	E	129	D	179	B
30	A	80	G	130	D	180	G
31	D	81	C	131	A	181	A
32	D	82	D	132	E	182	A
33	F	83	E	133	C	183	B
34	B	84	D	134	D	184	E
35	C	85	D	135	C	185	B
36	B	86	F	136	D	186	D
37	C	87	B	137	F	187	A
38	A	88	C	138	B	188	A
39	A	89	G	139	A	189	B
40	C	90	D	140	C	190	C
41	B	91	E	141	C	191	B
42	D	92	A	142	G	192	A
43	C	93	E	143	C	193	F
44	A	94	G	144	B	194	A
45	C	95	E	145	B	195	E
46	C	96	H	146	E	196	F
47	C	97	E	147	C	197	H
48	B	98	G	148	E	198	B
49	D	99	D	149	E	199	E
50	E	100	E	150	D	200	D

Question	Answer	Question	Answer	Question	Answer	Question	Answer
201	A	251	F	301	C	351	A
202	D	252	D	302	B	352	D
203	D	253	C	303	A	353	A
204	C	254	E	304	C	354	B
205	G	255	E	305	E	355	C
206	E	256	C	306	C	356	D
207	D	257	C	307	C	357	A
208	D	258	A	308	E	358	B
209	F	259	E	309	C	359	C
210	C	260	A	310	C	360	E
211	A	261	A	311	A		
212	B	262	C	312	D		
213	E	263	A	313	C		
214	C	264	F	314	B		
215	D	265	H	315	A		
216	C	266	C	316	C		
217	C	267	H	317	C		
218	E	268	B	318	E		
219	C	269	F	319	E		
220	B	270	E	320	A		
221	B	271	B	321	C		
222	B	272	A	322	C		
223	E	273	I	323	E		
224	D	274	C	324	A		
225	C	275	B	325	D		
226	A	276	D	326	C		
227	F	277	A	327	D		
228	F	278	B	328	A		
229	A	279	C	329	D		
230	C	280	C	330	B		
231	C	281	A	331	E		
232	D	282	B	332	A		
233	B	283	C	333	D		
234	A	284	B	334	C		
235	D	285	F	335	E		
236	D	286	E	336	E		
237	A	287	F	337	E		
238	C	288	A	338	C		
239	A	289	A	339	E		
240	E	290	A	340	D		
241	D	291	C	341	A		
242	A	292	A	342	B		
243	D	293	A	343	C		
244	D	294	F	344	D		
245	B	295	E	345	B		
246	A	296	D	346	D		
247	E	297	B	347	C		
248	D	298	D	348	E		
249	F	299	A	349	C		
250	F	300	F	350	A		

Section 1: Worked Answers

Question 1: B

Each three-block combination is mutually exclusive to any other combination, so the probabilities are added. Each block pick is independent of all other picks, so the probabilities can be multiplied. For this scenario there are three possible combinations:

P(2 red blocks and 1 yellow block) = P(red then red then yellow) + P(red then yellow then red) + P(yellow then red then red) =

$(\frac{12}{20} \times \frac{11}{19} \times \frac{8}{18}) + (\frac{12}{20} \times \frac{8}{19} \times \frac{11}{18}) + (\frac{8}{20} \times \frac{12}{19} \times \frac{11}{18}) =$

$\frac{3 \times 12 \times 11 \times 8}{20 \times 19 \times 18} = \frac{44}{95}$

Question 2: C

Multiply through by 15: $3(3x + 5) + 5(2x - 2) = 18 \times 15$

Thus: $9x + 15 + 10x - 10 = 270$

$9x + 10x = 270 - 15 + 10$

$19x = 265$

$x = 13.95$

Question 3: C

This is a rare case where you need to factorise a complex polynomial:

(3x   )(x   ) = 0, possible pairs: 2 x 10, 10 x 2, 4 x 5, 5 x 4

(3x - 4)(x + 5) = 0

3x - 4 = 0, so x = $\frac{4}{3}$

x + 5 = 0, so x = -5

Question 4: C

$\frac{5(x-4)}{(x+2)(x-4)} + \frac{3(x+2)}{(x+2)(x-4)}$

$= \frac{5x-20+3x+6}{(x+2)(x-4)}$

$= \frac{8x-14}{(x+2)(x-4)}$

Question 5: E

p α $\sqrt[3]{q}$, so p = k $\sqrt[3]{q}$

p = 12 when q = 27 gives 12 = k $\sqrt[3]{27}$, so 12 = 3k and k = 4

so p = 4 $\sqrt[3]{q}$

Now p = 24:

24 = 4$\sqrt[3]{q}$, so 6 = $\sqrt[3]{q}$ and q = 6^3 = 216

Question 6: A

8 x 9 = 72

8 = (4 x 2) = 2 x 2 x 2

9 = 3 x 3

$(2 \times 2 \times 2 \times 3 \times 3)^2 = 2 \times 2 \times 2 \times 2 \times 2 \times 2 \times 3 \times 3 \times 3 \times 3 = 2^6 \times 3^4$

Question 7: C

Note that 1.151 x 2 = 2.302.

Thus: $\frac{2 \times 10^5 + 2 \times 10^2}{10^{10}} = 2 \times 10^{-5} + 2 \times 10^{-8}$

$= 0.00002 + 0.00000002 = 0.00002002$

Question 8: E

$y^2 + ay + b$

$= (y + 2)^2 - 5 = y^2 + 4y + 4 - 5$

$= y^2 + 4y + 4 - 5 = y^2 + 4y - 1$

So a = 4 and y = -1

Question 9: E

Take $5(m + 4n)$ as a common factor to give: $\frac{4(m+4n)}{5(m+4n)} + \frac{5(m-2n)}{5(m+4n)}$

Simplify to give: $\frac{4m+16n+5m-10n}{5(m+4n)} = \frac{9m+6n}{5(m+4n)} = \frac{3(3m+2n)}{5(m+4n)}$

Question 10: C

$A \, \alpha \frac{1}{\sqrt{B}}$. Thus, $= \frac{k}{\sqrt{B}}$.

Substitute the values in to give: $4 = \frac{k}{\sqrt{25}}$.

Thus, $k = 20$.

Therefore, $A = \frac{20}{\sqrt{B}}$.

When B = 16, $A = \frac{20}{\sqrt{16}} = \frac{20}{4} = 5$

Question 11: E

Angles SVU and STU are opposites and add up to 180°, so STU = 91°

The angle of the centre of a circle is twice the angle at the circumference so SOU

$= 2 \times 91° = 182°$

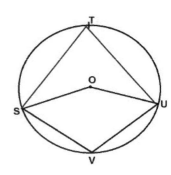

Question 12: E

The surface area of an open cylinder $A = 2\pi rh$. Cylinder B is an enlargement of A, so the increases in radius (r) and height (h) will be proportional: $\frac{r_A}{r_B} = \frac{h_A}{h_B}$. Let us call the proportion coefficient n, where $n = \frac{r_A}{r_B} = \frac{h_A}{h_B}$.

So $\frac{Area\ A}{Area\ B} = \frac{2\pi r_A h_A}{2\pi r_B h_B} = n\ x\ n = n^2$. $\frac{Area\ A}{Area\ B} = \frac{32\pi}{8\pi} = 4$, so n = 2.

The proportion coefficient n = 2 also applies to their volumes, where the third dimension (also radius, i.e. the r^2 in $V = \pi r^2 h$) is equally subject to this constant of proportionality. The cylinder's volumes are related by $n^3 = 8$.

If the smaller cylinder has volume 2π cm^3, then the larger will have volume $2\pi\ x\ n^3 = 2\pi\ x\ 8 = 16\pi$ cm^3.

Question 13: E

$$= \frac{8}{x(3-x)} - \frac{6(3-x)}{x(3-x)}$$

$$= \frac{8-18+6x}{x(3-x)}$$

$$= \frac{6x-10}{x(3-x)}$$

Question 14: B

For the black ball to be drawn in the last round, white balls must be drawn every round. Thus the probability is given by $P = \frac{9}{10}\ x\ \frac{8}{9}\ x\ \frac{7}{8}\ x\ \frac{6}{7}\ x\ \frac{5}{6}\ x\ \frac{4}{5}\ x\ \frac{3}{4}\ x\ \frac{2}{3}\ x\ \frac{1}{2}$

$$= \frac{9\ x\ 8\ x\ 7\ x\ 6\ x\ 5\ x\ 4\ x\ 3\ x\ 2\ x\ 1}{10\ x\ 9\ x\ 8\ x\ 7\ x\ 6\ x\ 5\ x\ 4\ x\ 3\ x\ 2\ x\ 1} = \frac{1}{10}$$

Question 15: C

The probability of getting a king the first time is $\frac{4}{52} = \frac{1}{13}$, and the probability of getting a king the second time is $\frac{3}{51}$. These are independent events, thus, the probability of drawing two kings is $\frac{1}{13}\ x\ \frac{3}{51} = \frac{3}{663} = \frac{1}{221}$

Question 16: B

The probabilities of all outcomes must sum to one, so if the probability of rolling a 1 is x, then: $x + x + x + x + 2x = 1$. Therefore, $x = \frac{1}{7}$.

The probability of obtaining two sixes $P_{12} = \frac{2}{7}\ x\ \frac{2}{7} = \frac{4}{49}$

Question 17: B

There are plenty of ways of counting, however the easiest is as follows: 0 is divisible by both 2 and 3. Half of the numbers from 1 to 36 are even (i.e. 18 of them). 3, 9, 15, 21, 27, 33 are the only numbers divisible by 3 that we've missed. There are 25 outcomes divisible by 2 or 3, out of 37.

Question 18: C

List the six ways of achieving this outcome: HHTT, HTHT, HTTH, TTHH, THTH, and THHT. There are 2^4 possible outcomes for 4 consecutive coin flips, so the probability of two heads and two tails is: $6\ x\ \frac{1}{2^4} = \frac{6}{16} = \frac{3}{8}$

Question 19: D

Count the number of ways to get a 5, 6 or 7 (draw the square if helpful). The ways to get a 5 are: 1, 4; 2, 3; 3, 2; 4, 1. The ways to get a 6 are: 1, 5; 2, 4; 3, 3; 4, 2; 5, 1. The ways to get a 7 are: 1, 6; 2, 5; 3, 4; 4, 3; 5, 2; 6, 1. That is 15 out of 36 possible outcomes.

	1	2	3	4	5	6
1	2	3	4	5	6	7
2	3	4	5	6	7	8
3	4	5	6	7	8	9
4	5	6	7	8	9	10
5	6	7	8	9	10	11
6	7	8	9	10	11	12

Question 20: C

There are x+y+z balls in the bag, and the probability of picking a red ball is $\frac{x}{(x+y+z)}$ and the probability of picking a green ball is $\frac{z}{(x+y+z)}$. These are independent events, so the probability of picking red then green is $\frac{xz}{(x+y+z)^2}$ and the probability of picking green then red is the same. These outcomes are mutually exclusive, so are added.

Question 21: B

There are two ways of doing it, pulling out a red ball then a blue ball, or pulling out a blue ball and then a red ball. Let us work out the probability of the first: $\frac{x}{(x+y+z)} \times \frac{y}{x+y+z-1}$, and the probability of the second option will be the same. These are mutually exclusive options, so the probabilities may be summed.

Question 22: A

[x: Player 1 wins point, y: Player 2 wins point]

Player 1 wins in five rounds if we get: yxxxx, xyxxx, xxyxx, xxxyx.

(Note the case of xxxxy would lead to player 1 winning in 4 rounds, which the question forbids.)

Each of these have a probability of $p^4(1-p)$. Thus, the solution is $4p^4(1-p)$.

Question 23: F

$4x + 7 + 18x + 20 = 14$

$22x + 27 = 14$

Thus, $22x = -13$

Giving $x = -\frac{13}{22}$

Question 24: D

$r^3 = \frac{3V}{4\pi}$

Thus, $r = \left(\frac{3V}{4\pi}\right)^{1/3}$

Therefore, $S = 4\pi \left[\left(\frac{3V}{4\pi}\right)^{\frac{1}{3}}\right]^2 = 4\pi \left(\frac{3V}{4\pi}\right)^{\frac{2}{3}}$

$= \frac{4\pi(3V)^{\frac{2}{3}}}{(4\pi)^{\frac{2}{3}}} = (3V)^{\frac{2}{3}} \times \frac{(4\pi)^1}{(4\pi)^{\frac{2}{3}}}$

$= (3V)^{\frac{2}{3}}(4\pi)^{1-\frac{2}{3}} = (4\pi)^{\frac{1}{3}}(3V)^{\frac{2}{3}}$

Question 25: A

Let each unit length be x.

Thus, $S = 6x^2$. Therefore, $x = \left(\frac{S}{6}\right)^{\frac{1}{2}}$

$V = x^3$. Thus, $V = [\left(\frac{S}{6}\right)^{\frac{1}{2}}]^3$ so $V = \left(\frac{S}{6}\right)^{\frac{3}{2}}$

Question 26: B

Multiplying the second equation by 2 we get $4x + 16y = 24$. Subtracting the first equation from this we get $13y = 17$, so $y = \frac{17}{13}$. Then solving for x we get $x = \frac{10}{13}$. You could also try substituting possible solutions one by one, although given that the equations are both linear and contain easy numbers, it is quicker to solve them algebraically.

Question 27: A

Multiply by the denominator to give: $(7x + 10) = (3y^2 + 2)(9x + 5)$

Partially expand brackets on right side: $(7x + 10) = 9x(3y^2 + 2) + 5(3y^2 + 2)$

Take x terms across to left side: $7x - 9x(3y^2 + 2) = 5(3y^2 + 2) - 10$

Take x outside the brackets: $x[7 - 9(3y^2 + 2)] = 5(3y^2 + 2) - 10$

Thus: $x = \frac{5(3y^2+ 2)-10}{7-9(3y^2+ 2)}$

Simplify to give: $x = \frac{(15y^2)}{(7 - 9(3y^2 + 2))}$

Question 28: F

$3x\left(\frac{3x^7}{x^{\frac{1}{3}}}\right)^3 = 3x\left(\frac{3^3 x^{21}}{x^{\frac{3}{3}}}\right)$

$= 3x\frac{27x^{21}}{x} = 81x^{21}$

Question 29: D

$2x[2^{\frac{7}{14}} x^{\frac{7}{14}}] = 2x[2^{\frac{1}{2}} x^{\frac{1}{2}}]$

$= 2x(\sqrt{2}\sqrt{x}) = 2\left[\sqrt{x}\sqrt{x}\right][\sqrt{2}\sqrt{x}]$

$= 2\sqrt{2x^3}$

Question 30: A

$A = \pi r^2$, therefore $10\pi = \pi r^2$

Thus, $r = \sqrt{10}$

Therefore, the circumference is $2\pi\sqrt{10}$

Question 31: D

$3.4 = 12 + (3 + 4) = 19$

$19.5 = 95 + (19 + 5) = 119$

Question 32: D

$$2.3 = \frac{2^3}{2} = 4$$

$$4.2 = \frac{4^2}{4} = 4$$

Question 33: F

This is a tricky question that requires you to know how to 'complete the square':

$$(x + 1.5)(x + 1.5) = x^2 + 3x + 2.25$$

Thus, $(x + 1.5)^2 - 7.25 = x^2 + 3x - 5 = 0$

Therefore, $(x + 1.5)^2 = 7.25 = \frac{29}{4}$

Thus, $x + 1.5 = \sqrt{\frac{29}{4}}$

Thus $x = -\frac{3}{2} \pm \sqrt{\frac{29}{4}} = -\frac{3}{2} \pm \frac{\sqrt{29}}{2}$

Question 34: B

Whilst you definitely need to solve this graphically, it is necessary to complete the square for the first equation to allow you to draw it more easily:

$$(x + 2)^2 = x^2 + 4x + 4$$

Thus, $y = (x + 2)^2 + 10 = x^2 + 4x + 14$

This is now an easy curve to draw (y = x² that has moved 2 units left and 10 units up). The turning point of this quadratic is to the left and well above anything in x³, so the only solution is the first intersection of the two curves in the upper right quadrant around (3.4, 39).

Question 35: C

By far the easiest way to solve this is to sketch them (don't waste time solving them algebraically). As soon as you've done this, it'll be very obvious that y = 2 and y = 1-x² don't intersect, since the latter has its turning point at (0, 1) and zero points at x = -1 and 1. y = x and y = x² intersect at the origin and (1, 1), and y = 2 runs through both.

Question 36: B

Notice that you're not required to get the actual values – just the number's magnitude. Thus, 897653 can be approximated to 900,000 and 0.009764 to 0.01. Therefore, 900,000 x 0.01 = 9,000

Question 37: C

Multiply through by 70: $7(7x + 3) + 10(3x + 1) = 14 \times 70$

Simplify: $49x + 21 + 30x + 10 = 980$

$79x + 31 = 980$

$x = \frac{949}{79}$

Question 38: A

Split the equilateral triangle into 2 right-angled triangles and apply Pythagoras' theorem:

$x^2 = \left(\frac{x}{2}\right)^2 + h^2$. Thus $h^2 = \frac{3}{4}x^2$

$h = \sqrt{\frac{3x^2}{4}} = \frac{\sqrt{3x^2}}{2}$

The area of a triangle = ½ x base x height $= \frac{1}{2}x\frac{\sqrt{3x^2}}{2}$

Simplifying gives: $x\frac{\sqrt{3x^2}}{4} = x\frac{\sqrt{3}\sqrt{x^2}}{4} = \frac{x^2\sqrt{3}}{4}$

Question 39: A

This is a question testing your ability to spot 'the difference between two squares'.

Factorise to give: $3 - \frac{7x(5x-1)(5x+1)}{(7x)^2(5x+1)}$

Cancel out: $3 - \frac{(5x-1)}{7x}$

Question 40: C

The easiest way to do this is to 'complete the square':

$(x-5)^2 = x^2 - 10x + 25$

Thus, $(x-5)^2 - 125 = x^2 - 10x - 100 = 0$

Therefore, $(x-5)^2 = 125$

$x - 5 = \pm\sqrt{125} = \pm\sqrt{25}\sqrt{5} = \pm5\sqrt{5}$

$x = 5 \pm 5\sqrt{5}$

Question 41: B

Factorise by completing the square:

$x^2 - 4x + 7 = (x-2)^2 + 3$

Simplify: $(x-2)^2 = y^3 + 2 - 3$

$x - 2 = \pm\sqrt{y^3 - 1}$

$x = 2 \pm \sqrt{y^3 - 1}$

Question 42: D

Square both sides to give: $(3x+2)^2 = 7x^2 + 2x + y$

Thus: $y = (3x+2)^2 - 7x^2 - 2x = (9x^2 + 12x + 4) - 7x^2 - 2x$

$y = 2x^2 + 10x + 4$

Question 43: C

This is a fourth order polynomial, which you aren't expected to be able to factorise at GCSE. This is where looking at the options makes your life a lot easier. In all of them, opening the bracket on the right side involves making $(y \pm 1)^4$ on the left side, i.e. the answers are hinting that $(y \pm 1)^4$ is the solution to the fourth order polynomial. Since there are negative terms in the equations (e.g. $-4y^3$), the solution has to be:

$(y-1)^4 = y^4 - 4y^3 + 6y^2 - 4y + 1$

Therefore, $(y-1)^4 + 1 = x^5 + 7$

Thus, $y - 1 = (x^5 + 6)^{\frac{1}{4}}$

$y = 1 + (x^5 + 6)^{1/4}$

Question 44: A

Let the width of the television be 4x and the height of the television be 3x.

Then by Pythagoras: $(4x)^2 + (3x)^2 = 50^2$

Simplify: $25x^2 = 2500$

Thus: $x = 10$. Therefore: the screen is 30 inches by 40 inches, i.e. the area is 1,200 inches^2.

Question 45: C

Square both sides to give: $1 + \frac{3}{x^2} = (y^5 + 1)^2$

Multiply out: $\frac{3}{x^2} = (y^{10} + 2y^5 + 1) - 1$

Thus: $x^2 = \frac{3}{y^{10} + 2y^5}$

Therefore: $x = \sqrt{\frac{3}{y^{10} + 2y^5}}$

Question 46: C

The easiest way is to double the first equation and triple the second to get:

$6x - 10y = 20 \text{ and } 6x + 6y = 39.$

Subtract the first from the second to give: $16y = 19$,

Therefore, $y = \frac{19}{16}$.

Substitute back into the first equation to give $x = \frac{85}{16}$.

Question 47: C

This is fairly straightforward; the first inequality is the easier one to work with: B and D and E violate it, so we just need to check A and C in the second inequality.

C: $1^3 - 2^2 < 3$, but A: $2^3 - 1^2 > 3$

Question 48: B

Whilst this can be done graphically, it's quicker to do algebraically (because the second equation is not as easy to sketch). Intersections occur where the curves have the same coordinates.

Thus: $x + 4 = 4x^2 + 5x + 5$

Simplify: $4x^2 + 4x + 1 = 0$

Factorise: $(2x + 1)(2x + 1) = 0$

Thus, the two graphs only intersect once at $x = -\frac{1}{2}$

Question 49: D

It's better to do this algebraically as the equations are easy to work with and you would need to sketch very accurately to get the answer. Intersections occur where the curves have the same coordinates. Thus: $x^3 = x$

$x^3 - x = 0$

Thus: $x(x^2 - 1) = 0$

Spot the 'difference between two squares': $x(x + 1)(x - 1) = 0$

Thus there are 3 intersections: at $x = 0, 1 \ and - 1$

Question 50: E

Note that the line is the hypotenuse of a right angled triangle with one side unit length and one side of length ½.

By Pythagoras, $\left(\frac{1}{2}\right)^2 + 1^2 = x^2$

Thus, $x^2 = \frac{1}{4} + 1 = \frac{5}{4}$

$x = \sqrt{\frac{5}{4}} = \frac{\sqrt{5}}{\sqrt{4}} = \frac{\sqrt{5}}{2}$

Question 51: D

We can eliminate z from equation (1) and (2) by multiplying equation (1) by 3 and adding it to equation (2):

3x + 3y − 3z = -3	Equation (1) multiplied by 3
2x − 2y +3z = 8	Equation (2) then add both equations
5x + y = 5	We label this as equation (4)

Now we must eliminate the same variable z from another pair of equations by using equation (1) and (3):

2x + 2y − 2z = -2	Equation (1) multiplied by 2
2x − y + 2z = 9	Equation (3) then add both equations
4x + y = 7	We label this as equation (5)

We now use both equations (4) and (5) to obtain the value of x:

5x + y = 5	Equation (4)
- 4x - y = -7	Equation (5) multiplied by -1
x = -2	

Substitute x back in to calculate y:

4x + y = 7

4(-2) + y = 7

- 8 + y = 7

y = 15

Substitute x and y back in to calculate z:

x + y − z = -1

-2 + 15 − z = -1

13 − z = -1

-z = -14

z = 14

Thus: x = -2, y = 15, z = 14

Question 52: D

This is one of the easier maths questions. Take 3a as a factor to give:

$3a(a^2 - 10a + 25) = 3a(a-5)(a-5) = 3a(a-5)^2$

Question 53: B

Note that 12 is the Lowest Common Multiple of 3 and 4. Thus:

-3 (4x + 3y) = -3 (48)   Multiply each side by -3

4 (3x + 2y) = 4 (34)   Multiply each side by 4

-12x – 9y = -144

<u>12x + 8y = 136</u>   Add together

-y = -8

y = 8

Substitute y back in: 4x + 3y = 48

4x + 3(8) = 48

4x + 24 = 48

4x = 24

x = 6

Question 54: E

Don't be fooled, this is an easy question, just obey BODMAS and don't skip steps.

$\frac{-(25-28)^2}{-36+14} = \frac{-(-3)^2}{-22}$

This gives: $\frac{-(9)}{-22} = \frac{9}{22}$

Question 55: E

Since there are 26 possible letters for each of the 3 letters in the license plate, and there are 10 possible numbers (0-9) for each of the 3 numbers in the same plate, then the number of license plates would be:

(26) x (26) x (26) x (10) x (10) x (10) = 17,576,000

Question 56: B

Expand the brackets to give: $4x^2 - 12x + 9 = 0$.

Factorise: $(2x - 3)(2x - 3) = 0$.

Thus, only one solution exists, x = 1.5.

Note that you could also use the fact that the discriminant, $b^2 - 4ac = 0$ to get the answer.

Question 57: C

$= \left(x^{\frac{1}{2}}\right)^{\frac{1}{2}} (y^{-3})^{\frac{1}{2}}$

$= x^{\frac{1}{4}} y^{-\frac{3}{2}} = \frac{x^{\frac{1}{4}}}{y^{\frac{3}{2}}}$

Question 58: A

Let x, y, and z represent the rent for the 1-bedroom, 2-bedroom, and 3-bedroom flats, respectively. We can write 3 different equations: 1 for the rent, 1 for the repairs, and the last one for the statement that the 3-bedroom unit costs twice as much as the 1-bedroom unit.

(1) x + y + z = 1240

(2) 0.1x + 0.2y + 0.3z = 276

(3) z = 2x

Substitute z = 2x in both of the two other equations to eliminate z:

(4) x + y + 2x = 3x + y = 1240

(5) 0.1x + 0.2y + 0.3(2x) = 0.7x + 0.2y = 276

-2(3x + y) = -2(1240)   Multiply each side of (4) by -2

10(0.7x + 0.2y) = 10(276)   Multiply each side of (5) by 10

(6) -6x -2y = -2480   Add these 2 equations

(7) 7x + 2y = 2760

x = 280

z = 2(280) = 560   Because z = 2x

280 + y + 560 = 1240   Because x + y + z = 1240

y = 400

Thus the units rent for £ 280, £ 400, £ 560 per week respectively.

Question 59: C

Following BODMAS:

$$= 5 \left[5(6^2 - 5 \times 3) + 400^{\frac{1}{2}} \right]^{1/3} + 7$$

$$= 5 \left[5(36 - 15) + 20 \right]^{\frac{1}{3}} + 7$$

$$= 5 \left[5(21) + 20 \right]^{\frac{1}{3}} + 7$$

$$= 5 (105 + 20)^{\frac{1}{3}} + 7$$

$$= 5 (125)^{\frac{1}{3}} + 7$$

$$= 5 (5) + 7$$

$$= 25 + 7 = 32$$

Question 60: B

Consider a triangle formed by joining the centre to two adjacent vertices. Six similar triangles can be made around the centre – thus, the central angle is 60 degrees. Since the two lines forming the triangle are of equal length, we have 6 identical equilateral triangles in the hexagon.

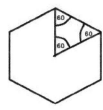

Now split the triangle in half and apply Pythagoras' theorem:

$$1^2 = 0.5^2 + h^2$$

Thus, $h = \sqrt{\frac{3}{4}} = \frac{\sqrt{3}}{2}$

Thus, the area of the triangle is: $\frac{1}{2}bh = \frac{1}{2} \times 1 \times \frac{\sqrt{3}}{2} = \frac{\sqrt{3}}{4}$

Therefore, the area of the hexagon is: $\frac{\sqrt{3}}{4} \times 6 = \frac{3\sqrt{3}}{2}$

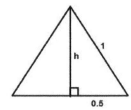

Question 61: B

Let x be the width and x+19 be the length.

Thus, the area of a rectangle is x(x + 19) = 780.

Therefore:

$x^2 + 19x - 780 = 0$

(x - 20)(x + 39) = 0

x − 20 = 0 or x + 39 = 0

x = 20 or x = -39

Since length can never be a negative number, we disregard x = -39 and use x = 20 instead.

Thus, the width is 20 metres and the length is 39 metres.

Question 62: B

The quickest way to solve is by trial and error, substituting the provided options. However, if you're keen to do this algebraically, you can do the following:

Start by setting up the equations: Perimeter = 2L + 2W = 34

Thus: L + W = 17

Using Pythagoras: $L^2 + W^2 = 13^2$

Since L + W = 17, W = 17 - L

Therefore: $L^2 + (17 - L)^2 = 169$

$L^2 + 289 - 34L + L^2 = 169$

$2L^2 - 34L + 120 = 0$

$L^2 - 17L + 60 = 0$

(L − 5) (L − 12) = 0

Thus: L = 5 and L = 12

And: W = 12 and W = 5

Question 63: C

Multiply both sides by 8: $4(3x - 5) + 2(x + 5) = 8(x + 1)$
Remove brackets: $12x - 20 + 2x + 10 = 8x + 8$
Simplify: $14x - 10 = 8x + 8$
Add 10: $14x = 8x + 18$
Subtract 8x: $6x = 18$
Therefore: $x = 3$

Question 64: C

Recognise that 1.742 x 3 is 5.226. Now, the original equation simplifies to: $= \frac{3 \times 10^6 + 3 \times 10^5}{10^{10}}$

$= 3 \times 10^{-4} + 3 \times 10^{-5} = 3.3 \times 10^{-4}$

Question 65: A

$Area = \frac{(2 + \sqrt{2})(4 - \sqrt{2})}{2}$

$= \frac{8 - 2\sqrt{2} + 4\sqrt{2} - 2}{2}$

$= \frac{6 + 2\sqrt{2}}{2}$

$= 3 + \sqrt{2}$

Question 66: C

Square both sides: $\frac{4}{x} + 9 = (y - 2)^2$

$\frac{4}{x} = (y - 2)^2 - 9$

Cross Multiply: $\frac{x}{4} = \frac{1}{(y-2)^2 - 9}$

$x = \frac{4}{y^2 - 4y + 4 - 9}$

Factorise: $x = \frac{4}{y^2 - 4y - 5}$

$x = \frac{4}{(y+1)(y-5)}$

Question 67: D

Set up the equation: $5x - 5 = 0.5 (6x + 2)$

$10x - 10 = 6x + 2$

$4x = 12$

$x = 3$

Question 68: C

Round numbers appropriately: $\frac{55 + (\frac{9}{4})^2}{\sqrt{900}} = \frac{55 + \frac{81}{16}}{30}$

81 rounds to 80 to give: $\frac{55 + 5}{30} = \frac{60}{30} = 2$

Question 69: D

There are three outcomes from choosing the type of cheese in the crust. For each of the additional toppings to possibly add, there are 2 outcomes: 1 to include and another not to include a certain topping, for each of the 7 toppings

Thus, the number of different kinds of pizza is: $3 \times 2 \times 2 \times 2 \times 2 \times 2 \times 2 \times 2 = 3 \times 2^7$

$= 3 \times 128 = 384$

Question 70: A

Although it is possible to do this algebraically, by far the easiest way is via trial and error. The clue that you shouldn't attempt it algebraically is the fact that rearranging the first equation to make x or y the subject leaves you with a difficult equation to work with (e.g. $x = \sqrt{1 - y^2}$) when you try to substitute in the second.

An exceptionally good student might notice that the equations are symmetric in x and y, i.e. the solution is when x = y. Thus $2x^2 = 1$ and $2x = \sqrt{2}$ which gives $\frac{\sqrt{2}}{2}$ as the answer.

Question 71: C

If two shapes are congruent, then they are the same size and shape. Thus, congruent objects can be rotations and mirror images of each other. The two triangles in E are indeed congruent (SAS). Congruent objects must, by definition, have the same angles.

Question 72: B

Rearrange the equation: $x^2 + x - 6 \geq 0$

Factorise: $(x + 3)(x - 2) \geq 0$

Remember that this is a quadratic inequality so requires a quick sketch to ensure you don't make a silly mistake with which way the sign is.
Thus, $y = 0$ when $x = 2$ and $x = -3$. $y > 0$ when $x > 2$ or $x < -3$.
Thus, the solution is: $x \leq -3 \; and \; x \geq 2$.

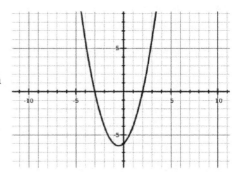

Question 73: B

Using Pythagoras: $a^2 + b^2 = x^2$

Since the triangle is isosceles: $a = b, \; so \; 2a^2 = x^2$

Area $= \frac{1}{2} base \; x \; height = \frac{1}{2}a^2$. From above, $a^2 = \frac{x^2}{2}$

Thus the area $= \frac{1}{2} x \frac{x^2}{2} = \frac{x^2}{4}$

Question 74: A

If X and Y are doubled, the value of Q increases by 4. Halving the value of A reduces this to 2. Finally, tripling the value of B reduces this to ⅔, i.e. the value decreases by ⅓.

Question 75: C

The quickest way to do this is to sketch the curves. This requires you to factorise both equations by completing the square:
$x^2 - 2x + 3 = (x - 1)^2 + 2$
$x^2 - 6x - 10 = (x - 3)^2 - 19$ Thus, the first equation has a turning point at (1, 2) and doesn't cross the x-axis. The second equation has a turning point at (3, -19) and crosses the x-axis twice.

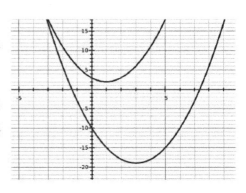

Question 76: F
That the amplitude of a wave determines its mass is false. Waves are not objects and do not have mass.

Question 77: A
We know that displacement s = 30 m, initial speed u = 0 ms^{-1}, acceleration a = 5.4 ms^{-2}, final speed v = ?, time t = ?
And that $v^2 = u^2 + 2as$
$v^2 = 0 + 2$ x 5.4 x 30
$v^2 = 324$ so v = 18 ms^{-1}
and s = ut + 1/2 at^2 so 30 = 1/2 x 5.4 x t^2
$t^2 = 30/2.7$ so t = 3.3 s

Question 78: D
The wavelength is given by: velocity v = λf and frequency f = 1/T so v = λ/T giving wavelength λ = vT
The period T = 49 s/7 so λ = 5 ms^{-1} x 7 s = 35 m

Question 79: E
This is a straightforward question as you only have to put the numbers into the equation (made harder by the numbers being hard to work with).
$Power = \frac{Force\ x\ Distance}{Time} = \frac{375\ N\ x\ 1.3\ m}{5\ s}$
$= 75\ x\ 1.3 = 97.5\ W$
Question 80: G
v = u + at
v = 0 + 5.6 x 8 = 44.8 ms^{-1}
And $s = ut + \frac{at^2}{2} = 0 + 5.6\ x\frac{8^2}{2} = 179.2$

Question 81: C
The sky diver leaves the plane and will accelerate until the air resistance equals their weight – this is their terminal velocity. The sky diver will accelerate under the force of gravity. If the air resistance force exceeded the force of gravity the sky diver would accelerate away from the ground, and if it was less than the force of gravity they would continue to accelerate toward the ground.

Question 82: D
 s = 20 m, u = 0 ms^{-1}, a = 10 ms^{-2}
and $v^2 = u^2 + 2as$
$v^2 = 0 + 2$ x 10 x 20
$v^2 = 400$;   v = 20 ms^{-1}
Momentum = Mass x velocity = 20 x 0.1 = 2 kgms^{-1}

Question 83: E
Electromagnetic waves have varying wavelengths and frequencies and their energy is proportional to their frequency.

Question 84: D
The total resistance = R + r = 0.8 + 1 = 1.8 Ω
and $I = \frac{e.m.f}{total\ resistance} = \frac{36}{1.8} = 20\ A$

Question 85: C
Use Newton's second law and remember to work in SI units:
So $Force = mass\ x\ accelaration = mass\ x\frac{\Delta velocity}{time}$
$= 20\ x\ 10^{-3}\ x\ \frac{100 - 0}{10\ x\ 10^{-3}}$
$= 200\ N$

Question 86: F
In this case, the work being done is moving the bag 0.7 m
i.e. $Work\ Done\ =\ Bag's\ Weight\ x\ Distance\ =\ 50\ x\ 10\ x\ 0.7 = 350\ N$
$Power = \frac{Work}{Time} = \frac{350}{3} = 116.7\ W$

$= 117\ W$ to 3 significant figures

Question 87: B
Firstly, use $P = Fv$ to calculate the power [Ignore the frictional force as we are not concerned with the resultant force here].
So $P = 300\ x\ 30 = 9000\ W$
Then, use $P = IV$ to calculate the current.
$I = P/V = 9000/200 = 45\ A$

Question 88: C
Work is defined as $W = F\ x\ s$. Work can also be defined as work = force x distance moved in the direction of force. Work is measured in joules and 1 Joule = 1 Newton x 1 Metre, and 1 Newton = 1 Kg x ms^{-2} [F = ma].
Thus, 1 Joule = Kgm^2s^{-2}

Question 89: G
Joules are the unit of energy (and also Work = Force x Distance). Thus, 1 Joule = 1 N x 1 m.
Pa is the unit of Pressure (= Force/Area). Thus, Pa = N x m^{-2}. So J = Nm^{-2} x m^3 = Pa x m^3. Newton's third law describes that every action produces an equal and opposite reaction. For this reason, the energy required to decelerate a body is equal to the amount of energy it possess during movement, i.e. its kinetic energy, which is defined as in statement 1.

Question 90: D
Alpha radiation is of the lower energy, as it represents the movement of a fairly large particle consisting of 2 neutrons and 2 protons. Beta radiation consists of high-energy, high-speed electrons or positrons.

Question 91: E
The half-life does depend on atom type and isotope, as these parameters significantly impact on the physical properties of the atom in general, so statement 1 is false. Statement 2 is the correct definition of half-life. Statement 3 is also correct: half-life in exponential decay will always have the same duration, independent of the quantity of the matter in question; in non-exponential decay, half-life is dependent on the quantity of matter in question.

Question 92: A
In contrast to nuclear fission, where neutrons are shot at unstable atoms, nuclear fusion is based on the high speed, high-temperature collision of molecules, most commonly hydrogen, to form a new, stable atom while releasing energy.

Question 93: E
Nuclear fission releases a significant amount of energy, which is the basis of many nuclear weapons. Shooting neutrons at unstable atoms destabilises the nuclei which in turn leads to a chain reaction and fission. Nuclear fission can lead to the release of ionizing gamma radiation.

Question 94: G
The total resistance of the circuit would be twice the resistance of one resistor and proportional to the voltage, as given by Ohm's Law. Since it is a series circuit, the same current flows through each resistor and since they are identical the potential difference across each resistor will be the same.

Question 95: E
The distance between Earth and Sun = Time x Speed = 60 x 8 seconds x 3 x 10^8 ms^{-1} = 480 x 3 x 10^8 m
Approximately = 1500 x 10^8 = 1.5 x 10^{11} m.
The circumference of Earth's orbit around the sun is given by $2\pi r$ = 2 x 3 x 1.5 x 10^{11}
= 9 x 10^{11} = 10^{12} m

Question 96: H
Speed is a scalar quantity whilst velocity is a vector describing both magnitude and direction. Speed describes the distance a moving object covers over time (i.e. speed = distance/time), whereas velocity describes the rate of change of the displacement of an object (i.e. velocity = displacement/time). The internationally standardised unit for speed is meters per second (ms^{-1}), while ms^{-2} is the unit of acceleration.

Question 97: E
Ohm's Law only applies to conductors and can be mathematically expressed as $V \alpha I$. The easiest way to do this is to write down the equations for statements c, d and e. C: $I \alpha \frac{1}{V}$; D: $I \alpha V^2$; E: $I \alpha V$. Thus, statement E is correct.

Question 98: G
Any object at rest is not accelerating and therefore has no resultant force. Strictly speaking, Newton's second law is actually: Force = rate of change of momentum, which can be mathematically manipulated to give statement 2:
$$Force = \frac{momentum}{time} = \frac{mass \times velocity}{time} = mass \times accelaration$$

Question 99: D
Statement 3 is incorrect, as $Charge = Current \times time$. Statement 1 substitutes $I = \frac{V}{R}$ and statement 2 substitutes $I = \frac{P}{V}$.

Question 100: E
Weight of elevator + people = mg = 10 x (1600 + 200) = 18,000 N
Applying Newton's second law of motion on the car gives:
Thus, the resultant force is given by:
F$_M$ = Motor Force – [Frictional Force + Weight]
F$_M$ = M – 4,000 – 18,000
Use Newton's second law to give: F$_M$ = M – 22,000 N = ma
Thus, M – 22,000 N = 1,800a
Since the lift must accelerate at 1ms^{-2}: M = 1,800 kg x 1 ms^{-2} + 22,000 N
M = 23,800 N

Question 101: D
Total Distance = Distance during acceleration phase + Distance during braking phase
Distance during <u>acceleration phase</u> is given by:
$$s = ut + \frac{at^2}{2} = 0 + \frac{5 \times 10^2}{2} = 250 \ m$$
$$v = u + at = 0 + 5 \times 10 = 50 \ ms^{-1}$$
And use $a = \frac{v-u}{t}$ to calculate the deceleration: $a = \frac{0-50}{20} = -2.5 \ ms^{-2}$
Distance during the <u>deceleration phase</u> is given by:
$$s = ut + \frac{at^2}{2} = 50 \times 20 + \frac{-2.5 \times 20^2}{2} = 1000 - \frac{2.5 \times 400}{2}$$
$$s = 1000 - 500 = 500 \ m$$
Thus, $Total \ Distance = 250 + 500 = 750 \ m$

Question 102: G
It is not possible to calculate the power of the heater as we don't know the current that flows through it or its internal resistance. The 8 ohms refers to the external copper wire and not the heater. Whilst it's important that you know how to use equations like P = IV, it's more important that you know when you *can't* use them!

Question 103: F
This question has a lot of numbers but not any information on time, which is necessary to calculate power. You cannot calculate power by using P= IV as you don't know how many electrons are accelerated through the potential difference per unit time. Thus, more information is required to calculate the power.

Question 104: B
When an object is in equilibrium with its surroundings, it radiates and absorbs energy at the same rate and so its temperature remains constant i.e. there is no *net* energy transfer. Radiation is slower than conduction and convection.

Question 105: A
The work done by the force is given by: $Work\ Done = Force \times Distance = 12\,N \times 3\,m = 36\,J$

Since the surface is frictionless, $Work\ Done = Kinetic\ Energy.$
$E_k = \frac{mv^2}{2} = \frac{6v^2}{2}$
Thus, $36 = 3v^2$
$v = \sqrt{12} = \sqrt{4}\sqrt{3} = 2\sqrt{3}\ ms^{-1}$

Question 106: C
$Total\ energy\ supplied\ to\ water = Change\ in\ temperature \times Mass\ of\ water \times 4,000\,J$
$= 40 \times 1.5 \times 4,000 = 240,000\,J$

$Power\ of\ the\ heater = \frac{Work\ Done}{time} = \frac{240,000}{50 \times 60} = \frac{240,000}{3,000} = 80\,W$

Using $P = IV = \frac{V^2}{R}$:

$R = \frac{V^2}{P} = \frac{100^2}{80} = \frac{10,000}{80} = 125\ ohms$

Question 107: G
The large amount of energy released during atomic fission is the basis underlying nuclear power plants. Splitting an atom into two or more parts will by definition produce molecules of different sizes than the original atom; therefore it produces two new atoms. The free neutrons and photons produced by the splitting of atoms form the basis of the energy release.

Question 108: D
Gravitational potential energy is just an extension of the equation work done = force x distance (force is the weight of the object, *mg*, and distance is the height, *h*). The reservoir in statement 3 would have a potential energy of 10^{10} Joules i.e. 10 Giga Joules ($E_p = 10^6$ kg x 10 N x 10^3 m).

Question 109: D
Statement 1 is the common formulation of Newton's third law. Statement 2 presents a consequence of the application of Newton's third law.

Statement 3 is false: rockets can still accelerate because the products of burning fuel are ejected in the opposite direction from which the rocket needs to accelerate.

Question 110: E
Positively charged objects have lost electrons. $Charge = Current \times Time = \frac{Voltage}{Resistance} \times Time.$
Objects can become charged by friction as electrons are transferred from one object to the other.

Question 111: B
Each body of mass exerts a gravitational force on another body with mass. This is true for all planets as well. Gravitational force is dependent on the mass of both objects. Satellites stay in orbit due to centripetal force that acts tangentially to gravity (not because of the thrust from their engines). Two objects will only land at the same time if they also have the same shape or they are in a vaccum (as otherwise air resistance would result in different terminal velocities).

Question 112: A
Metals conduct electrical charge easily and provide little resistance to the flow of electrons. Charge can also flow in several directions. However, all conductors have an internal resistance and therefore provide *some* resistance to electrical charge.

Question 113: E

First, calculate the rate of petrol consumption:

$$\frac{Speed}{Consumption} = \frac{60 \ miles/hour}{30 \ miles/gallon} = 2 \ gallons/hour$$

Therefore, the total power is:

$2 \ gallons = 2 \ x \ 9 \ x \ 10^8 = 18 \ x \ 10^8 J$

$1 \ hour = 60 \ x \ 60 = 3600 \ s$

$\text{Power} = \frac{Energy}{Time} = \frac{18 \ x 10^8}{3600}$

$P = \frac{18}{36} \ x \ 10^6 = 5 \ x \ 10^5 \ W$

Since efficiency is 20%, the power delivered to the wheels $= 5 \ x \ 10^5 \ x \ 0.2 = 10^5 \ W = 100 \ kW$

Question 114: D

Beta radiation is stopped by a few millimetres of aluminium, but not by paper. In β^- radiation, a neutron changes into a proton plus an emitted electron. This means the atomic mass number remains unchanged.

Question 115: F

Firstly, calculate the mass of the car $= \frac{Weight}{g} = \frac{15,000}{10} = 1,500 \ kg$

Then using $v = u + at$ where v = 0 ms^{-1} and u = 15 ms^{-1} and t = 10 x 10^{-3} s

$a = \frac{0-15}{0.01} = 1500 ms^{-2}$

$F = ma = 1500 \ x \ 1500 = 2 \ 250 \ 000 \ N$

Question 116: E

Electrical insulators offer high resistance to the flow of charge. Insulators are usually non-metals; metals conduct charge very easily. Since charge does not flow easily to even out, they can be charged with friction.

Question 117: A

The car accelerates for the first 10 seconds at a constant rate and then decelerates after t=30 seconds. It does not reverse, as the velocity is not negative. Therefore only statement 1 is not true.

Question 118: B

The distance travelled by the car is represented by the area under the curve (integral of velocity) which is given by the area of two triangles and a rectangle:

$Area = \left(\frac{1}{2} \ x \ 10 \ x \ 10\right) + (20 \ x \ 10) + \left(\frac{1}{2} \ x \ 10 \ x \ 10\right)$

$Area = 50 + 200 + 50 = 300 \ m$

Question 119: C

Using the equation force = mass x acceleration, where the unknown acceleration = change in velocity over change in time.

Hence: $\frac{F}{m} = \frac{change \ in \ velocity}{change \ in \ time}$

We know that F = 10,000 N, mass = 1,000 kg and change in time is 5 seconds.

So, $\frac{10,000}{1,000} = \frac{change \ in \ velocity}{5}$

So change in velocity $= 10 \ x \ 5 = 50 \ m/s$

Question 120: D

This question tests both your ability to convert unusual units into SI units and to select the relevant values (e.g. the crane's mass is not important here).

0.01 tonnes = 10 kg; 100 cm = 1 m; 5,000 ms = 5 s

$$Power = \frac{Work\ Done}{Time} = \frac{Force\ x\ Distance}{Time}$$

In this case the force is the weight of the wardrobe $= 10 \times g = 10 \times 10 = 100N$

Thus, $Power = \frac{100\ x\ 1}{5} = 20\ W$

Question 121: F

Remember that the resistance of a parallel circuit (R_T) is given by: $\frac{1}{R_T} = \frac{1}{R_1} + \frac{1}{R_2} + \ ...$

Thus, $\frac{1}{R_T} = \frac{1}{1} + \frac{1}{2} = \frac{3}{2}$ and therefore $R = \frac{2}{3}\ \Omega$

Using Ohm's Law: $I = \frac{20\ V}{\frac{2}{3}\Omega} = 20 \times \frac{3}{2} = 30\ A$

Question 122: E

Water is denser than air. Therefore, the speed of light decreases when it enters water and increases when it leaves water. The direction of light also changes when light enters/leaves water. This phenomenon is known as refraction and is governed by Snell's Law.

Question 123: C

The voltage in a parallel circuit is the same across each branch, i.e. branch A Voltage = branch B Voltage.
The resistance of Branch $A = 6 \times 5 = 30\ \Omega$; the resistance of Branch $B = 10 \times 2 = 20\ \Omega$.
Using Ohm's Law: I= V/R. Thus, $I_A = \frac{60}{30} = 2\ A$; $I_B = \frac{60}{20} = 3\ A$

Question 124: C

This is a very straightforward question made harder by the awkward units you have to work with. Ensure you are able to work comfortably with prefixes of 10^9 and 10^{-9} and convert without difficulty.
50,000,000,000 nano Watts = 50 W and 0.000000004 Giga Amperes = 4 A.
Using $P = IV$: $V = \frac{P}{I} = \frac{50}{4} = 12.5\ V = 0.0125\ kV$

Question 125: B

Radioactive decay is highly random and unpredictable. Only gamma decay releases gamma rays and few types of decay release X-rays. The electrical charge of an atom's nucleus decreases after alpha decay as two protons are lost.

Question 126: D

Using $P = IV$: $I = \frac{P}{V} = \frac{60}{15} = 4\ A$
Now using Ohm's Law: $R = \frac{V}{I} = \frac{15}{4} = 3.75\ \Omega$
So each resistor has a resistance of $\frac{3.75}{3} = 1.25\ \Omega$.
If two more resistors are added, the overall resistance $= 1.25 \times 5 = 6.25\ \Omega$

Question 127: F

To calculate the useful work done and hence the efficiency, we must know the resistive forces on the tractor, whether it is stationery or moving at the end point and if there is any change in vertical height.

Question 128: G

Electromagnetic induction is defined by statements 1 and 2. An electrical current is generated when a coil moves in a magnetic field.

Question 129: D

An ammeter will always give the same reading in a series circuit, not in a parallel circuit where current splits at each branch in accordance with Ohm's Law.

Question 130: D
Electrons move in the opposite direction to current (i.e. they move from negative to positive).

Question 131: A
For a fixed resistor, the current is directly proportional to the potential difference. For a filament lamp, as current increases, the metal filament becomes hotter. This causes the metal atoms to vibrate and move more, resulting in more collisions with the flow of electrons. This makes it harder for the electrons to move through the lamp and results in increased resistance. Therefore, the graph's gradient decreases as current increases.

Question 132: E
Vector quantities consist of both direction and magnitude, e.g. velocity, displacement, etc., and can be added by taking account of direction in the sum.

Question 133: C
The gravity on the moon is 6 times less than 10 ms^{-2}. Thus, $g_{moon}= \frac{10}{6} = \frac{5}{3}$ ms^{-2}.

Since weight = mass x gravity, the mass of the rock $= \frac{250}{\frac{5}{3}} = \frac{750}{5} = 150 \ kg$

Therefore, the density $= \frac{mass}{volume} = \frac{150}{250} = 0.6 \ kg/cm^3$

Question 134: D
An alpha particle consists of a helium nucleus. Thus, alpha decay causes the mass number to decrease by 4 and the atomic number to decrease by 2. Five iterations of this would decrease the mass number by 20 and the atomic number by 10.

Question 135: C
Using Ohm's Law: The potential difference entering the transformer (V_1) = 10 x 20 = 200 V

Now use $\frac{N1}{N2} = \frac{V1}{V2}$ to give: $\frac{5}{10} = \frac{200}{V2}$

Thus, $V_2 = \frac{2,000}{5} = 400$ V

Question 136: D
For objects in free fall that have reached terminal velocity, acceleration = 0.
Thus, the sphere's weight = resistive forces.
Using Work Done = Force x Distance: Force = 10,000 J/100 m = 100 N.
Therefore, the sphere's weight = 100 N and since $g = 10$ms^{-2}, the sphere's mass = 10 kg

Question 137: F
The wave length of ultraviolet waves is longer than that of x-rays. Wavelength is inversely proportional to frequency. Most electromagnetic waves are not stopped with aluminium (and require thick lead to stop them), and they travel at the speed of light. Humans can only see a very small part of the spectrum.

Question 138: B
If an object moves towards the sensor, the wavelength will appear to decrease and the frequency increase. The faster this happens, the faster the increase in frequency and decrease in wavelength.

Question 139: A
$Acceleration = \frac{Change \ in \ Velocity}{Time} = \frac{1,000}{0.1} = 10,000 \ ms^{-2}$

Using Newton's second law: The Braking Force = Mass x Acceleration.

Thus, Braking Force = 10,000 x 0.005 $= 50 \ N$

Question 140: C
Polonium has undergone alpha decay. Thus, Y is a helium nucleus and contains 2 protons and 2 neutrons. Therefore, 10 moles of Y contain $2 \times 10 \times 6 \times 10^{23}$ protons $= 120 \times 10^{23} = 1.2 \times 10^{25}$ protons.

Question 141: C
The rod's activity is less than 1,000 Bq after 300 days. In order to calculate the longest possible half-life, we must assume that the activity is just below 1,000 Bq after 300 days. Thus, the half-life has decreased activity from 16,000 Bq to 1,000 Bq in 300 days.
After one half-life: Activity = 8,000 Bq
After two half-lives: Activity = 4,000 Bq
After three half-lives: Activity = 2,000 Bq
After four half-lives: Activity = 1,000 Bq
Thus, the rod has halved its activity a minimum of 4 times in 300 days. 300/4 = 75 days

Question 142: G
There is no change in the atomic mass or proton numbers in gamma radiation. In β decay, a neutron is transformed into a proton (and an electron is released). This results in an increase in proton number by 1 but no overall change in atomic mass. Thus, after 5 rounds of beta decay, the proton number will be $89 + 5 = 94$ and the mass number will remain at 200. Therefore, there are 94 protons and $200-94 = 106$ neutrons.
NB: You are not expected to know about β^+ decay.

Question 143: C
Calculate the speed of the sound $= \frac{distance}{time} = \frac{500}{1.5} = 333 \, ms^{-1}$

Thus, the $Wavelength = \frac{Speed}{Frequency} = \frac{333}{440}$

Approximate 333 to 330 to give: $\frac{330}{440} = \frac{3}{4} = 0.75 \, m$

Question 144: B
Firstly, note the all the answer options are a magnitude of 10 apart. Thus, you don't have to worry about getting the correct numbers as long as you get the correct power of 10. You can therefore make your life easier by rounding, e.g. approximate π to 3, etc.
The area of the shell $= \pi r^2$.
$= \pi \times (50 \times 10^{-3})^2 = \pi \times (5 \times 10^{-2})^2$
$= \pi \times 25 \times 10^{-4} = 7.5 \times 10^{-3} \, m^2$
The deceleration of the shell $= \frac{u-v}{t} = \frac{200}{500 \times 10^{-6}} = 0.4 \times 10^6 \, ms^{-2}$
Then, using Newton's Second Law: $Braking \, force = mass \times acceleration = 1 \times 0.4 \times 10^6 = 4 \times 10^5 N$
Finally: $Pressure = \frac{Force}{Area} = \frac{4 \times 10^5}{7.5 \times 10^{-3}} = \frac{8}{15} \times 10^8 \, Pa \approx 5 \times 10^7 Pa$

Question 145: B
The fountain transfers 10% of 1,000 J of energy per second into 120 litres of water per minute. Thus, it transfers 100 J into 2 litres of water per second.
Therefore the Total Gravitational Potential Energy, $E_p = mg\Delta h$
Thus, $100 \, J = 2 \times 10 \times h$
Hence, $h = \frac{100}{20} = 5 \, m$

Question 146: E
In step down transformers, the number of turns of the primary coil is larger than that of the secondary coil to decrease the voltage. If a transformer is 100% efficient, the electrical power input = electrical power output (P=IV).

Question 147: C
The percentage of C^{14} in the bone halves every 5,730 years. Since it has decreased from 100% to 6.25%, it has undergone 4 half-lives. Thus, the bone is $4 \times 5,730$ years old = 22,920 years

Question 148: E
This is a straightforward question in principle, as it just requires you to plug the **values into the equation:**
$Velocity = Wavelength\ x\ Frequency$ – Just ensure you work in SI units to get the **correct answer.**
$Frequency = \frac{2\ m/s}{2.5\ m} = 0.8\ Hz = 0.8\ x\ 10^{-6} MHz = 8\ x\ 10^{-7}\ MHz$

Question 149: E
If an element has a half-life of 25 days, its BQ value will be halved every 25 days.
A total of 350/25 = 14 half-lives have elapsed. Thus, the count rate has halved 14 times. **Therefore, to calculate** the original rate, the final count rate must be doubled 14 times = $50\ x\ 2^{14}$.
$2^{14} = 2^5\ x\ 2^5\ x\ 2^4 = 32\ x\ 32\ x\ 16 = 16,384.$
Therefore, the original count rate = $16,384\ x\ 50 = 819,200$

Question 150: D
Remember that $V = IR = \frac{P}{I}$ and $Power = \frac{Work\ Done}{Time} = \frac{Force\ x\ Distance}{Time} = Force\ x\ Velocity$;

Thus, **A** is derived from: $V = IR$,

B is derived from: $= \frac{P}{I}$,

C is derived from: $Voltage = \frac{Power}{Current} = \frac{Force\ x\ Velocity}{Current}$,

Since $Charge = Current\ x\ Time$, **E** and **F** are derived from: $Voltage = \frac{Power}{Current} = \frac{Force\ x\ Distance}{Time\ x\ Current} = \frac{J}{As} = \frac{J}{C}$,

D is incorrect as Nm = J. Thus the correct variant would be NmC^{-1}

Question 151: D
Different isotopes are differentiated by the number of neutrons in the core. This gives them **different molecular weights** and different chemical properties with regards to stability. The number of protons **defines each element,** and the number of electrons its charge.

Question 152: E
A displacement reaction occurs when a more reactive element displaces a less reactive **element in its compound.** All 4 reactions are examples of displacement reactions as a less reactive element is **being replaced by a more** reactive one.

Question 153: A
There needs to be 3Ca, 12H, 14O and 2P on each side. Only option A satisfies this.

Question 154: A
To balance the equation there needs to be 9Ag, 9N, 9O_3, 9K, 3P on each side. Only option A **satisfies this.**

Question 155: D
A more reactive halogen can displace a less reactive halogen. Thus, chlorine can displace **bromine and iodine from** an aqueous solution of its salts, and fluorine can replace chlorine. The trend is the opposite **for alkali metals, where** reactivity increases down the group as electrons are further from the core and easier to **lose.**

Question 156: C
$2Mg + O_2 = 2MgO$
so $2\ x\ 24 = 48$ and $2\ x\ (24 + 16) = 80$
so 48 g of magnesium produces 80g of magnesium oxide
so 1g of magnesium produces 1g x 80g/48g = 1.666g oxide
so 75g x 1.666 = 125g

Question 157: B
$H_2 + 2OH^- \rightarrow 2H_2O + e^-$
Thus, the hydrogen loses electrons i.e. is oxidised.

Question 158: F
Ammonia is 1 nitrogen and 3 hydrogen atoms bonded covalently. N = 14g and H = 1g per mole, so percentage of N in NH_3 = 14g/17g = 82%. It can be produced from N_2 through fixation or the industrial Haber process for use in fertiliser, and may break down to its components.

Question 159: A
Milk is weakly acidic, pH 6.5-7.0, and contains fat. This is broken down by lipase to form fatty acids - turning the solution slightly more acidic.

Question 160: C
Glucose loses four hydrogen atoms; one definition of an oxidation reaction is a reaction in which there is loss of hydrogen.

Question 161: C
Isotopes have the same number of protons and electrons, but a different number of neutrons. The number of neutrons has no impact on the rate of reactions.

Question 162: E
$Mg + H_2SO_4 \rightarrow MgSO_4 + H_2$
Number of moles of Mg $= \frac{6}{24} = 0.25$ moles.
1 mole of Mg reacts with 1 mole H_2SO_4 to produce 1 mole of magnesium sulphate. Therefore, 0.25 moles H_2SO_4 will react to produce 0.25 moles of $MgSO_4$.
M_r of $H_2SO_4 = 2 + 32 + 64 = 98g$ per mole
The mass of H_2SO_4 used = 0.25 moles x 98g per mole = 24.5g.

Since 30g of H_2SO_4 is present, H_2SO_4 is in excess and the magnesium is the limiting reagent.

M_r of $MgSO_4 = 24 + 32 + 64 = 120g$ per mole
The mass of $MgSO_4$ produced = 0.25 moles x 120g per mole = 30g which is the same mass as that of sulphuric acid in the original reaction.

Question 163: F
Reactivity series of metals:
Cu is more reactive than Ag and will displace it.
Ca is more reactive than H and will displace it.
2 and 4 are incorrect because Fe is higher in the reactivity series than Cu and Fe is lower in the reactivity series than Ca, so no displacement will occur.

Question 164: G
Moving left to right is the equivalent of moving down the metal reactivity series (i.e. Na is most reactive and Zn is least reactive). Therefore, moving from left to right, the reactivity of the metals decreases, likelihood of corrosion decreases, less energy is required to separate metals from their ores and metals lose electrons less readily to form positive ions.

Question 165: F
Halogens become less reactive as you progress down group 17. Thus in order of increasing reactivity from left to right: I→ Br→ Cl. Therefore, I will not displace Br, Cl will displace Br and Br will displace I.

Question 166: A
Wires are made out of copper because it is a good conductor of electricity. Copper is also used in coins (not aluminium). Aluminium is resistant to corrosion but because of a layer of aluminium oxide (not hydroxide).

Question 167: C
$2Li + 2H_2O \rightarrow 2LiOH + H_2$
Therefore, 2 moles of Li react to produce 1 mole of H_2 gas (24 dm^3).
The number of moles of Li $= \frac{21}{7} = 3$ moles.
Thus, 1.5 moles of H_2 gas are produced = 36 dm^3.

Question 168: B

$MgCl_2$ contains stronger bonds than NaCl because Mg ions have a 2+ charge, thus having a stronger electrostatic pull for negative chloride ions. The smaller atomic radius also means that the nucleus has less distance between it and incoming electrons. Transition metals are able to form multiple stable ions e.g. Fe^{2+} and Fe^{3+}.

Covalently bonded structures do tend to have lower MPs than ionically bonded, but the giant covalent structures (diamond and graphite for example) have very high melting points. Graphite is an example of a covalently bonded structure which conducts electricity.

Question 169: D

Energy is released from reaction **A**, as shown by a negative enthalpy. The reaction is therefore exothermic. Since energy is released, the product CO_2 has less energy than the reactants did. Therefore, CO_2 is more stable. Reaction **B** has a positive enthalpy, which means energy must be put into the reaction for it to occur i.e. it's an endothermic reaction. That means that the products (CaO and CO_2) have more energy and are less stable than the reactants ($CaCO_3$).

Question 170: B

Solid oxides are unable to conduct electricity because the ions are immobile. Metals are extracted from their molten ores by electrolysis. Fractional distillation is used to separate miscible liquids with similar boiling points. Mg^{2+} ions have a greater positive charge and a smaller ionic radius than Na^+ ions, and therefore have stronger bonds.

Question 171: E

Li^+ (2) and Na^+ (2, 8)
Mg^{2+} (2, 8) and Ne (2, 8)
Na^{2+} (2, 7) and Ne (2, 8)
O^{2-} (2, 4) and a Carbon atom (2, 4)

Question 172: B

Reactivity of both group 1 and 2 increases as you go down the groups because the valence electrons that react are further away from the positively charged nucleus (which means the electrostatic attraction between them is weaker). Group 1 metals are usually more reactive because they only need to donate one electron, whilst group 2 metals must donate two electrons.

Question 173: D

This is a straightforward question that tests basic understanding of kinetics. Catalysts help overcome energy barriers by reducing the activation energy necessary for a reaction.

Question 174: D

H^1 contains 1 proton and no neutrons. Isotopes have the same numbers of protons, but different numbers of neutrons. Thus, H^3 contains two more neutrons than H^1.

Question 175: D

Oxidation is the loss of electrons and reduction is the gain of electrons (therefore increasing electron density). Halogens tend to act as electron recipients in reactions and are therefore good oxidising agents.

Question 176: D

These statements all come from the Kinetic Theory of Gases, an idealised model of gases that allows for the derivation of the ideal gas law. The angle at which gas molecules move is not related to temperature; movement is random. Gas molecules lose no energy when they collide with each other, collisions are assumed elastic. The average kinetic energy of gas molecules is the same for all gases at the same temperature as they are assumed to be point masses. Momentum = mass x velocity. Therefore, the momentum of gas molecules increases with pressure as a greater force is exerted on each molecule.

Question 177: E

An exothermic reaction is defined as a chemical reaction that releases energy. Thus, aerobic respiration producing life energy, the burning of magnesium, and the reacting of acids/bases are almost always exothermic processes. Similarly, the combustion of most things (including hydrogen) is exothermic. Evaporation of water is a physical process in which no chemical reaction is taking place.

Question 178: E

$2 C_3H_6 + 9 O_2 \rightarrow 6 H_2O + 6 CO_2$
Assign the oxidation numbers for each element:
For C_3H_6: C = -2; H = +1
For O_2: O = 0
For H_2O: H = +1; O = -2
For CO_2: C = +4; O = -2
Look for the changes in the oxidation numbers:
H remained at +1
C changed from -2 to +4. Thus, it was oxidized
O changed from 0 to -2. Thus, it was reduced.

Question 179: B
The equation for the reaction is: $Zn + CuSO_4 \rightarrow ZnSO_4 + Cu$
Assign oxidation numbers for each element:
For Zn: Zn = 0
For $CuSO_4$: Cu = +2; S = +6; O = -2
For $ZnSO_4$: Zn = +2; S = +6; O = -2
For Cu: Cu = 0
With these oxidation numbers, we can see that Zn was oxidized and Cu in $CuSO_4$ was reduced. Thus, Zn acted as the reducing agent and Cu in $CuSO_4$ is the oxidizing agent.

Question 180: B
Acids are proton donors which only exist in aqueous solution, which is a liquid state. Strong acids are fully ionised in solution and the reaction between an acid and a base $\rightarrow$ salt + water.
The pH of weak acids is usually between 4 and 6.

Question 181: G
Let x be the relative abundance of Z^6 and y the relative abundance of Z^8.
The average atomic mass takes the abundances of all 3 isotopes into account.
Thus, (Abundance of Z^5)(Mass Z^5) + (Abundance of Z^6)(Mass Z^6) + (Abundance of Z^8)(Mass Z^8) = 7
Therefore: (5 x 0.2) + 6x + 8y = 7
So: 6x + 8y = 6
Divide by two to give: 3x + 4y = 3
The abundances of all isotopes = 100% = 1
This gives: 0.2 + x + y = 1
Solve the two equations simultaneously:
y = 0.8 – x
3x + 4(0.8 – x) = 3
3x + 3.2 – 4x = 3
Therefore, x = 0.2
y = 0.8 - 0.2 = 0.6
Thus, the overall abundances are Z^5 = 20%, Z^6 = 20% and Z^8 = 60%. Therefore, all the statements are correct.

Question 182: A
If a metal is more reactive than hydrogen, a displacement reaction will occur resulting in the formation of a salt with the metal cation and hydrogen.

Question 183: B

$6\ FeSO_4 + K_2Cr_2O_7 + 7\ H_2SO_4 \rightarrow 3\ (Fe)_2(SO_4)_3 + Cr_2(SO_4)_3 + K_2SO_4 + 7\ H_2O$

In order to save time, you have to quickly eliminate options (rather than try every combination out). The quickest way is to do this is algebraically:

For Potassium:

$2b = 2e = 2f$

Therefore, $b = f$.

Option F does not fulfil $b = e = f$.

For Iron:

$a = 2d$

Options C, D and E don't fulfil $a = 2d$.

For Hydrogen:

$2c = 2g$

Therefore, $c = g$.

Option A does not fulfil $c = g$.

This leaves option B as the answer.

Question 184: E

Atoms are electrically neutral. Ions have different numbers of electrons when compared to atoms of the same element. Protons provide just under 50% of an atom's mass, the other 50% is provided by neutrons. Isotopes don't exhibit significantly different kinetics. Protons do indeed repel each other in the nucleus (which is one reason why neutrons are needed: to reduce the electrical charge density).

Question 185: B

The noble gasses are extremely useful, e.g. helium in blimps, neon signs, argon in bulbs. They are colourless and odourless and have no valence electrons. As with the rest of the periodic table, boiling point increases as you progress down the group (because of increased Van der Waals forces). Helium is the most abundant noble gas (and indeed the 2nd most abundant element in the universe).

Question 186: D

Alkenes can be hydrogenated (i.e. reduced) to alkanes. Aromatic compounds are commonly written as cyclic alkenes, but their properties differ from those of alkenes. Therefore alkenes and aromatic compounds do not belong to the same chemical class.

Question 187: A

The average atomic mass takes the abundances of both isotopes into account:

(Abundance of Cl^{35})(Mass Cl^{35}) + (Abundance of Cl^{37})(Mass Cl^{37}) = 35.453

34.969(Abundance of Cl^{35}) + 36.966(Abundance of Cl^{37}) = 35.453

The abundances of both isotopes = 100% = 1

I.e. abundance of Cl^{35} + abundance of Cl^{37} = 1

Therefore: $x + y = 1$ which can be rearranged to give: $y = 1-x$

Therefore: $x + (1 - x) = 1$.

$34.969x + 36.966(1-x) = 35.453$

$x = 0.758$

$1 - x = 0.242$

Therefore, Cl^{35} is 3 times more abundant than Cl^{37}.

Note that you could approximate the values here to arrive at the solution even quicker, e.g. 34.969 $\rightarrow$ 35, 36.966 $\rightarrow$ 37 and 35.453 $\rightarrow$ 35.5

Question 188: A

Transition metals form multiple stable ions which may have many different colours (e.g. green Fe^{2+} and brown Fe^{3+}). They usually form ionic bonds and are commonly used as catalysts (e.g. iron in the Haber process, Nickel in alkene hydrogenation). They are excellent conductors of electricity and are known as the d-block elements.

Question 189: B

$2Na + 2H_2O \rightarrow 2NaOH + H_2$

$8000 \text{ cm}^3 = 8 \text{ dm}^3 = \frac{1}{3}$ moles of H_2

2 moles of Na react completely to form 1 mole of H_2.

Therefore, $\frac{2}{3}$ moles of Na must have reacted to produce $\frac{1}{3}$ moles of Hydrogen. $\frac{2}{3}$ x 23g per mole = 15.3g.

% Purity of sample $= \frac{15.3}{20}$ x 100 = 76.5%

Question 190: C

Assume total mass of molecule is 100g. Therefore, it contains 70.6g carbon, 5.9g hydrogen and 23.5g oxygen.

Now, calculate the number of moles of each element using $Moles = \frac{Mass}{Molar\ Mass}$

$Moles\ of\ Carbon = \dfrac{70.6}{12} \approx 6$

$Moles\ of\ Hydrogen = \dfrac{5.9}{1} \approx 6$

$Moles\ of\ Oxygen = \dfrac{23.5}{16} \approx 1.5$

Therefore, the molar ratios give an empirical formula of $C_6H_6O_{1.5} = C_4H_4O$.

Molar mass of the empirical formula = (4 x 12) + (4 x 1) + 16 = 68.

Molar mass of chemical formula = 136. Therefore, the chemical formula = $C_8H_8O_2$.

Question 191: B

$S + 6 HNO_3 \rightarrow H_2SO_4 + 6 NO_2 + 2 H_2O$

In order to save time, you have to quickly eliminate options (rather than try every combination out).

The quickest way to do this is algebraically:

For Hydrogen:

$b = 2c + 2e$

Options A, C, D, E and F don't fulfil $b = 2c + 2e$.

This leaves options B as the only possible answer.

Note how quickly we were able to get the correct answer here by choosing an element that appears in 3 molecules (as opposed to Sulphur or Nitrogen which only appear in 2).

Question 192: A

Alkenes undergo addition reactions, such as that with hydrogen, when catalysed by nickel, whilst alkanes do not as they are already fully saturated. The C=C bond is stronger than the C-C bond, but it is not exactly twice as strong, so will not require twice the energy to break it. Both molecules are organic and will dissolve in organic solvents.

Question 193: F

Diamond is unable to conduct electricity because all the electrons are involved in covalent bonds. Graphite is insoluble in water + organic solvents. Graphite is also able to conduct electricity because there are free electrons that are not involved in covalent bonds.

Methane and Ammonia both have low melting points. Methane is not a polar molecule, so cannot conduct electricity or dissolve in water. Ammonia is polar and will dissolve in water. It can conduct electricity in aqueous form, but not as a gas.

Question 194: A

Catalysts increase the rate of reaction by providing an alternative reaction path with a lower activation energy, which means that less energy is required and so costs are reduced. The point of equilibrium, the nature of the products, and the overall energy change are unaffected by catalysts.

Question 195: E

The 5 carbon atoms in this hydrocarbon make it a "pent" stem. The C=C bond makes it an alkene, and the location of this bond is the 2nd position, making the molecule pent-2-ene.

Question 196: F
Group 1 elements form positively charged ions in most reactions and therefore lose electrons. Thus, the oxidation number must increase. Their reactivity increases as the valence electrons are further away from the positively charged nucleus down group. Whilst they all react **spontaneously** with oxygen, only the latter half of Group 1 elements react **instantaneously**.

Question 197: H
The cathode attracts positively charged ions. The cathode reduces ions and the anode oxidises ions. Electrolysis can be used to separate compounds but not mixtures (i.e. substances that are not chemically joined).

Question 198: B
Pentane, C_5H_{12}, has a total of 3 isomers. A, C and D are correctly configured. However, the 4th Carbon atom in option B has more than 4 bonds which wouldn't be possible. If you're stuck on this – draw them out!

Question 199: E
$3\ Cu + 8\ HNO_3 \rightarrow 3\ Cu(NO_3)_2 + 2\ NO + 4\ H_2O$
In order to save time, you have to quickly eliminate options (rather than try every combination out).
The quickest way to do this is algebraically, by first assigning coefficients to the equation:
$\mathbf{a}Cu + \mathbf{b}HNO_3 \rightarrow \mathbf{c}Cu(NO_3)_2 + \mathbf{d}NO + \mathbf{e}H_2O$
For Nitrogen: b = 2c + d.
In this case, only option E satisfies b = 2c + d.
Note that using copper wouldn't be as useful, as all the options satisfy a = c.

Question 200: D
Alkenes are an organic series and have twice as many hydrogen atoms as carbon atoms. Bromine water is decolourised in their presence and they take part in addition reactions. Alkenes are more reactive than alkanes because they contain a C=C bond.

Question 201: A
Group 17 elements are missing one valence electron, so form negative ions. Bromine is a liquid at room temperature, and is also coloured brown. Reactivity decreases as you progress down Group 17, so fluorine reacts more vigorously than iodine. All Group 17 elements are found bound to each other, e.g. F_2 and Cl_2.

Question 202: D
CO poisoning and spontaneous combustion do not occur in the electrolysis of brine. The products of cathode and anode in the electrolysis of brine are Cl_2 and H_2. If these two gases react with each other they can form HCl, which is extremely corrosive.

Question 203: D
The hydrogen produced is positively charged and therefore needs to be reduced by the addition of an electron before being released. This happens at the cathode. The chlorine produced is negatively charged and therefore needs to lose electrons. This happens at the anode. NaOH is formed in this process.

Question 204: C
Alkanes are made of chains of singly bonded carbon and hydrogen atoms. C-H bonds are very strong and confer alkanes a great deal of stability. An alkane with 14 hydrogen atoms is called Hexane, as it has 6 carbon atoms. Alkanes burn in excess oxygen to produce carbon dioxide and water. Bromine water is decolourised in the presence of alkenes.

Question 205: G
You've probably got a lot of experience of organic chemistry by now, so this should be fairly straightforward. Alcohols by definition contain an R-OH functional group and because of this polar group are highly soluble in water. Ethanol is a common biofuel.

Question 206: E
Alkanes are saturated (and therefore non-reducible), have the general formula C_nH_{2n+2} and have no effect on Bromine solution. Alkenes are unsaturated (and therefore reducible), have the general formula C_nH_{2n} and turn bromine water colourless because they can undergo an addition reaction with bromine.

Question 207: D

The balanced equation for the reaction between magnesium oxide and hydrochloric acid is:

$O + 2HCl \rightarrow MgCl_2 + H_2$

The relative molecular mass of MgO is $24 + 16 = 40$g per mole.

Therefore 10g of MgO represents $10/40 = 0.25$ moles.

As the ratio of MgO to $MgCl_2$ is 1:1, we know that the amount of $MgCl_2$ produced will also be 0.25 moles. One mole of $MgCl_2$ has a molecular mass of $24 + (2 \times 35.5) = 95$g per mole.

Therefore the reaction will produce $0.25 \times 95 = 23.75$g of $MgCl_2$.

Question 208: D

Moving up the alkane series, as size and mass of the molecule increases (and thus the Van der Waals forces increase), the boiling point and viscosity increase and the flammability and volatility decrease. Therefore pentadecane will be more viscous than pentane.

Question 209: F

All of the factors mentioned will affect the rate of a reaction. The temperature affects the movement rate of particles, which if moving faster in higher temperatures will collide more often, thus increasing the rate of reaction. Collision rate is also increased with a higher concentration of reactants, and with a higher concentration of a catalyst or one with larger surface area, which will provide more active sites, thus increasing the rate of reaction.

Question 210: C

The total atomic mass of the end product is $C[12 + (2 \times 16)] + D[(2 \times 1) + 16] = 44C + 18D$

We know that $176 = 44C$. Therefore $C = 4$, and that $108 = 18D$ so $D = 6$.

Thus, the equation becomes: $C_aH_b + O_2 \rightarrow 4CO_2 + 6H_2O$.

This gives a ratio of 4C to 12H, which is a ratio of 1:3 carbon to hydrogen. This means the unknown hydrocarbon must be a multiple of this ratio. By balancing the equation we can see that the unknown hydrocarbon must be ethane, C_2H_6: $2C_2H_6 + 7O_2 \rightarrow 4CO_2 + 6H_2O$.

Question 211: A

$C_2H_5OH \rightarrow C_2H_4O$. Thus, ethanol has lost two hydrogen atoms, i.e. has been oxidised. Note that although another substrate may be reduced (therefore making it a redox reaction), ethanol has only been oxidised.

Question 212: B

This is fairly straightforward but you can save time by doing it algebraically:

For Barium: $3a = b$

For Nitrogen: $2a = c$

Let $a = 1$, thus, $b = 3$ and $c = 2$

Question 213: E

There are 14 oxygen atoms on the left side. Thus: $3b + 2c = 14$.

Note also that for Sulphur: $a = c$, and for Iron: $a = 2b$.

This sets up an easy trio of simultaneous equations:

Substitute a into the first equation to give: $1.5a + 2a = 14$. Thus: $a = 14/3.5 = 4$.

Therefore, $a = c = 4$ and $b = 2$

Question 214: C

The average atomic mass takes the abundances of all isotopes into account:

Mass = (Abundance of Mg^{23})(Mass Mg^{23}) + (Abundance of Mg^{25})(Mass Mg^{25}) + (Abundance of Mg^{26})(Mass Mg^{26})

$Mass = 23 \times 0.80 + 25 \times 0.10 + 26 \times 0.10$

$= 18.4 + 2.5 + 2.6 = 23.5$

Question 215: D

Cl_2 and Fe_2O_3 are reduced in their reactions and are therefore oxidising agents. Similarly, CO and Cu^{2+} are oxidised in their reactions and are therefore reducing agents. Cl is a stronger oxidising agent than Br as it is higher up in the reactivity series, and will displace negative Br ions from its compounds to form the oxidised Br_2. Mg is a stronger reducing agent than Cu, as it is higher up in the reactivity series. Thus, Mg would displace a positive copper ion from its compound to form copper atoms. Therefore Mg reduces Cu.

Question 216: C

NaCl is an ionic compound and therefore has a high melting point. It is highly soluble in water but only conducts electricity in solution/as a liquid.

Question 217: C

The equation for the reaction is: $2NaOH + Zn(NO_3)_2 \rightarrow 2NaNO_3 + Zn(OH)_2$
Therefore, the molar ratio between NaOH and $Zn(OH)_2$ is 2:1.
Molecular Mass of NaOH = 23 + 16 + 1 = 40
Molecular Mass of $Zn(OH)_2$ = 65 + 17 x 2 = 99
Thus, the number of moles of NaOH that react = 80/40 = 2 moles.
Therefore, 1 mole of $Zn(OH)_2$ is produced. Mass = 99g per mole x 1 mole = 99g

Question 218: E

Metal + Water → Hydroxide + Hydrogen gas; the reaction is always exothermic. Reactivity increases down the group, so potassium reacts more vigorously with water than sodium. Therefore all are correct.

Question 219: C

Electrolysis separates NaCl into sodium and chloride ions but not CO_2 (which is a covalently bound gas). Sieves cannot separate ionically bound compounds like NaCl. Dyes are miscible liquids and can be separated by chromatography. Oil and water are immiscible liquids, so a separating funnel is necessary to separate the mixtures. Methane and diesel are separated from each other during fractional distillation, as they have different boiling points.

Question 220: B

The reaction between water and caesium can cause spontaneous combustion and therefore doesn't make the reaction safer. The reaction between caesium and fluoride is highly exothermic and does not require a catalyst. The reaction produces CsF which is a salt.

Question 221: B

The nucleus of larger elements contain more neutrons than protons to reduce the charge density, e.g. Br^{80} contains 35 protons but 45 neutrons. Stable isotopes very rarely undergo radioactive decay.

Question 222: B

The vast majority of salts contain ionic bonds that require a significant amount of heat energy to break.

Question 223: E

306ml of water is 306g, which is the equivalent of 306g/18g per mole of H_2O = 17 moles. 17 times Avogadro's constant gives the number of molecules present, which is 1.02×10^{25}. There are 10 protons and 10 electrons in each water molecule. Hence there are 1.02×10^{26} protons.

Question 224: D

The number of moles of each element = Mass/Molar Mass. Let the % represent the mass in grams: Hydrogen: 3.45g/1g per mole = 3.45 moles
Oxygen: 55.2g/16g per mole = 3.45 moles
Carbon: 41.4g/12g per mole = 3.45 moles
Thus, the molar ratio is 1:1:1. The only option that satisfies this is option D.

Question 225: C

Group 17 elements are non-metals, whilst group 2 elements are metals. Thus, the Group 17 element must gain electrons when it reacts with the Group 2 element, i.e. B is reduced. The easy way to calculate the formula is to swap the valences of both elements: A is +2 and B is -1. Thus, the compound is AB_2.

Question 226: A

DNA consists of 4 bases: adenine, guanine, thymine and cysteine. The sugar backbone consists of deoxyribose, hence the name DNA. DNA is found in the cytoplasm of prokaryotes.

Question 227: F

Mitochondria are responsible for energy production by ATP synthesis. Animal cells do not have a cell wall, only a cell membrane. The endoplasmic reticulum is important in protein synthesis, as this is where the proteins are assembled.

Question 228: F

If you aren't studying A-level biology, this question may stretch you. However, it is possible to reach an answer by process of elimination. Mitochondria are the 'powerhouse' of the cell in aerobic respiration, responsible for cell energy production rather than DNA replication or protein synthesis. As energy producers they are required in muscle cells in large numbers, and in sperm cells to drive the tail responsible for movement. They are enveloped by a double membrane, possibly because they started out as independent prokaryotes engulfed by eukaryotic cells.

Question 229: A

The majority of bacteria are commensals and don't lead to disease.

Question 230: C

Bacteria carry genetic information on plasmids and not in nuclei like animal cells. They don't need meiosis for replication, as they do not require gametes. Bacterial genomes consist of DNA, just like animal cells.

Question 231: C

Active transport requires a transport protein and ATP, as work is being done against an electrochemical gradient. Unlike diffusion, the relative concentrations of the materials being transported aren't important.

Question 232: D

Meiosis produces haploid gametes. This allows for fusion of 2 gametes to reach a full diploid set of chromosomes again in the zygote.

Question 233: B

Mendelian inheritance separates traits into dominant or recessive. It applies to all sexually reproducing organisms. Don't get confused by statement C – the offspring of 2 heterozygotes has a 25% chance of expressing a recessive trait, but it will be homozygous recessive.

Question 234: A

Hormones are released into the bloodstream and act on receptors in different organs in order to cause relatively slow changes to the body's physiology. Hormones frequently interact with the nervous system, e.g. Adrenaline and Insulin, however, they don't directly cause muscles to contract. Almost all hormones are synthesised.

Question 235: D

Neuronal signalling can happen via direct electrical stimulation of nerves or via chemical stimulation of synapses which produces a current that travels along the nerves. Electrical synapses are very rare in mammals, the majority of mammalian synapses are chemical.

Question 236: D

Remember that pH changes cause changes in electrical charge on proteins (= polypeptides) that could interfere with protein – protein interactions. Whilst the other statements are all correct to a certain extent, they are the downstream effects of what would happen if enzymes (which are also proteins) didn't work.

Question 237: A

The bacterial cell wall is made up of cellulose and protects the bacterium from the external environment, in particular from osmotic stresses, and is important in most bacteria.

Question 238: C

Sexual reproduction relies on formation of gametes during **meiosis**. Mitosis doesn't produce genetically distinct cells. Mitosis is, however, the basis for tissue growth.

Question 239: A

A mutation is a permanent change in the nucleotide sequence of DNA. Whilst mutations may lead to changes in organelles and chromosomes, or even be harmful, they are strictly defined as permanent changes to the DNA or RNA sequence.

Question 240: E

Mutations are fairly common, but in the vast majority of cases do not have any impact on phenotype due to the redundancy of the genome. Sometimes they can confer selective advantages and allow organisms to survive better (i.e. evolve by natural selection), or they can lead to cancers as cells start dividing uncontrollably.

Question 241: D

Antibodies represent a pivotal molecule of the immune system. They provide very pointed and selective targeting of pathogens and toxins without causing damage to the body's own cells.

Question 242: A

Kidneys are not involved in digestion, but do filter the blood of waste products. Glucose is found in high concentrations in the urine of diabetics, who cannot absorb it without working insulin.

Question 243: D

Hormones are slower acting than nerves and act for a longer time. Hormones also act in a more general way. Adrenaline is also a hormone released into the body causing the fight-or-flight response. Although it is quick acting, it still lasts for a longer time than a nervous response, as you can still feel its effects for a time after the response, e.g. shaking hands.

Question 244: D

Homeostasis is about minimising changes to the internal environment by modulating both input and output.

Question 245: B

There is less energy and biomass each time you move up a trophic level. Only 10% of consumed energy is transferred to the next trophic level, so only one tenth of the previous biomass can be sustained in the next trophic level up.

Question 246: A

In asexual reproduction, there is no fusion of gametes as the single parent cell divides. There is therefore no mixing of chromosomes and, as a result, no genetic variation.

Question 247: E

The image is first formed on the retina which conveys it to the brain via a sensory nerve. The brain then sends an impulse to the muscle via a motor neuron.

Question 248: D

Blood from the kidney returns to the heart via the renal (kidney-related) vein, which drains into the inferior vena cava. The blood then passes through the pulmonary vasculature (veins carry blood to the heart, arteries away from the heart) before going into the aorta and eventually the hepatic (liver-related) artery.

Question 249: F

Clones are genetically identical by definition, and a large number of them could conceivably reduce the gene pool of a population. In adult cell cloning, the genetic material of an egg is replaced with the genetic material of an adult cell. Cloning is possible for all DNA based life forms, including plants and other types of animals.

Question 250: F

Gene varieties cause intraspecies variation, e.g. different eye colours. If mutations confer a selective advantage, those individuals with the mutation will survive to reproduce and grow in numbers. Genetic variation is caused by mixing of parent genomes and mutations. Species with similar characteristics often do have similar genes.

Question 251: F

Alleles are different versions of the same gene. If you are a homozygous for a trait, you have two identical alleles for that particular gene, and if you are heterozygous you have two different alleles of that gene. Recessive traits only appear in the phenotype when there are no dominant alleles for that trait, i.e. two recessive alleles are carried.

Question 252: D

Remember that red blood cells don't have a nucleus and therefore have no DNA. In meiosis, a diploid cell divides in such a way so as to produce four haploid cells. Any type of cell division will require energy.

Question 253: C

The hypothalamus detects too little water in the blood, so the pituary gland releases ADH. The kidney maintains the blood water level, and allows less water to be lost in the urine until the blood water level returns to normal.

Question 254: E

Venous blood has a higher level of carbon dioxide and lower oxygen. Carbon dioxide forms carbonic acid in aqueous solution, thus making the pH of venous blood slightly more acidic than arterial blood. This leaves only E and F as possibilities, but releasing pH levels cannot fluctuate significantly gives pH 7.4.

Question 255: E

The cytoplasm is 80% water, but also contains, among other things, electrolytes and proteins. The cytoplasm doesn't contain everything, e.g. DNA is found in the nucleus.

Question 256: C

ATP is produced in mitochondria in aerobic respiration and in the cytoplasm during anaerobic respiration only.

Question 257: C

The cell membrane allows both active transport and passive transport by diffusion of certain ions and molecules, and is found in eukaryotes and prokaryotes like bacteria. It is a phospholipid bilayer.

Question 258: A

1 and 2 only: 223 PAIRS = 446 chromosomes; meiosis produces 4 daughter cells with half of the original number of chromosomes each, while mitosis produces two daughter cells with the original number of chromosomes each.

Question 259: E

If Bob is homozygous dominant (RR) the probability of having a child with red hair is 0%. However, if Bob is heterozygous (Rr), there is a 50% chance of having a child with red hair, since Mary must be homozygous recessive (rr) to have red hair. As we do not know Bob's genotype, both possibilities must be considered.

Question 260 A

If an offspring is born with red hair, it confirms Bob is heterozygous (Rr). He cannot have a red-haired child if he is homozygous dominant (RR), and would himself have red hair were he homozygous recessive (rr).

Question 261: A

Monohybrid cross rr and Rr results in 50% Rr and 50% rr offspring. 50% of offspring will have black hair, but they will be heterozygous for the hair allele.

Question 262: C

When the chest walls expand, the intra-thoracic pressure decreases. This causes the atmospheric pressure outside the chest to be greater than pressure inside the chest, resulting in a flow of air into the chest.

Question 263: A

Producers are found at the bottom of food chains and always have the largest biomass.

Question 264: F
All the statements are true; the carbon and nitrogen cycles are examinable in Section 2, so make sure you understand them! The atmosphere is 79% inert N_2 gas, which must be 'fixed' to useable forms by high-energy lightning strikes or by bacterial mediation. Humans also manually fix nitrogen for fertilisers with the Haber process.

Question 265: H
None of the above statements are correct. Mutations can be silent, cause a loss of function, or even a gain in function, depending on the exact location in the gene and the base affected. Mutations only cause a change in protein structure if the amino acids expressed by the gene affected are changed. This is normally due to a shift in reading frame. Whilst cancer arises as a result of a series of mutations, very few mutations actually lead to cancer.

Question 266: C
Remember that heart rate is controlled via the autonomic nervous system, which isn't a part of the central nervous system.

Question 267: H
None of the above are correct. There is no voluntary input to the heart in the form of a neuronal connection. Parasympathetic neurones slow the heart and sympathetic nervous input accelerates heart rate.

Question 268: B
If lipase is not working, fat from the diet will not be broken down, and will build up in the stool. Lactase, for instance, is responsible for breaking down lactose, and its malfunctioning causes lactose-intolerance.

Question 269: F
Oxygenated blood flows from the lungs to the heart via the pulmonary vein. The pulmonary artery carries deoxygenated blood from the heart to the lungs. Animals like fish have single circulatory systems. Deoxygenated blood is found in the superior vena cava, returning to the heart from the body. Veins in the arms and hands frequently don't have valves.

Question 270: E
Enzymatic digestion takes place throughout the GI tract, including in the mouth (e.g. amylase), stomach (e.g. pepsin), and small intestine (e.g. trypsin). The large intestine is primarily responsible for water absorption, whilst the rectum acts as a temporary store for faecal matter (i.e. digestion has finished by the rectum).

Question 271: B
This is an example of the monosynaptic stretch reflex; these reflexes are performed at the spinal level and therefore don't involve the brain.

Question 272: A
Statement 2 describes diffusion, as CO_2 is moving with the concentration gradient. Statement 3 describes active transport, as amino acids are moving against the concentration gradient.

Question 273: I
3 is the correct equation for animals, and 4 is correct for plants.

Question 274: C
The mitochondria are only the site for aerobic respiration, as anaerobic respiration occurs in the cytoplasm. Aerobic respiration produces more ATP per substrate than anaerobic respiration, and therefore is also more efficient. The chemical equation for glucose being respired aerobically is: $C_6H_{12}O_6 + 6O_2 \rightarrow 6CO_2 + 6H_2O$. Thus, the molar ratio is 1:6 (i.e. each mole glucose produces 6 moles of CO_2).

Question 275: B
The nucleus contains the DNA and chromosomes of the cell. The cytoplasm contains enzymes, salts and amino acids in addition to water. The plasma membrane is a bilayer. Lastly, the cell wall is indeed responsible for protecting vs. increased osmotic pressures.

Question 276: D

When a medium is hypertonic relative to the cell cytoplasm, it is more concentrated than the cytoplasm, and when it is hypotonic, it is less concentrated. So, when a medium is hypotonic relative to the cell cytoplasm, the cell will gain water through osmosis. When the medium is isotonic, there will be no net movement of water across the cell membrane. Lastly, when the medium is hypertonic relative to the cell cytoplasm, the cell will lose water by osmosis.

Question 277: A

Stem cells have the ability to differentiate and produce other kinds of cells. However, they also have the ability to generate cells of their own kind and stem cells are able to maintain their undifferentiated state. The two types of stem cells are embryonic stem cells and adult stem cells. The adult stem cells are present in both children and adults.

Question 278: B

All of the following statements are examples of natural selection, except for the breeding of horses. Breeding and animal husbandry are notable methods of artificial selection, which are brought about by humans.

Question 279: C

Enzymes create a stable environment to stabilise the transition state. Enzymes do not distort substrates. Enzymes generally have little effect on temperature directly. Lastly, they are able to provide alternative pathways for reactions to occur.

Question 280: C

A negative feedback system seeks to minimise changes in a system by modulating the response in accordance with the error that's generated. Salivating before a meal is an example of a feed-forward system (i.e. salivating is an anticipatory response). Throwing a dart does not involve any feedback (during the action). pH and blood pressure are both important homeostatic variables that are controlled via powerful negative feedback mechanisms, e.g. massive haemorrhage leads to compensatory tachycardia.

Question 281: A

One of the major functions of white blood cells is to defend the body against bacterial and fungal infections. They can kill pathogens by engulfing them and also use antibodies to help them recognise pathogens. Antibodies are produced by white blood cells.

Question 282: B

The CV system does indeed transport nutrients and hormones. It also increases blood flow to exercising muscles (via differential vasodilatation) and also helps with thermoregulation (e.g. vasoconstriction in response to cold). The respiratory system is responsible for oxygenating blood.

Question 283: C

Adrenaline always increases heart rate and is almost always released during sympathetic responses. It travels primarily in the blood and affects multiple organ systems. It is also a potent vasoconstrictor.

Question 284: B

Protein synthesis occurs in the cytoplasm. Proteins are usually coded by several amino acids. Red blood cells lack a nucleus and, therefore, the DNA to create new proteins. Protein synthesis is a key part of mitosis, as it allows the parent cell to grow prior to division.

Question 285: F

Remember that most enzymes work better in neutral environments (amylase works even better at slightly alkaline pH). Thus, adding sodium bicarbonate will increase the pH and hence increase the rate of activity. Adding carbohydrate will have no effect, as the enzyme is already saturated. Adding amylase will increase the amount of carbohydrate that can be converted per unit time. Increasing the temperature to $100°$ C will denature the enzyme and reduce the rate.

Question 286: E

Taking the normal allele to be C and the diseased allele to be c, one can model the scenario with the following Punnett square:

		Carrier Mother	
		C	c
Diseased Father	c	Cc	cc
	c	Cc	cc

The gender of the children is irrelevant as the inheritance is autosomal recessive, but we see that all children produced would inherit at least one diseased allele.

Question 287: F

All of the organs listed have endocrine functions. The thyroid produces thyroid hormone. The ovary produces oestrogen. The pancreas secretes glucagon and insulin. The adrenal gland secretes adrenaline. The testes produce testosterone.

Question 288: A

Insulin works to decrease blood glucose levels. Glucagon causes blood glucose levels to increase; glycogen is a carbohydrate. Adrenaline works to increase heart rate.

Question 289: A

The left side of the heart contains oxygenated blood from the lungs which will be pumped to the body. The right side of the heart contains deoxygenated blood from the body to be pumped to the lungs.

Question 290: A

Since Individual 1 is homozygous normal, and individual 5 is heterozygous and affected, the disease must be dominant. Since males only have one X-chromosome, they cannot be carriers for X-linked conditions. If Nafram syndrome was X-linked, then parents 5 and 6 would produce sons who always have no disease and daughters that always do. As this is not the case shown in individuals 7-10, the disease must be autosomal dominant.

Question 291: C

We know that the inheritance of Nafram syndrome is autosomal dominant, so using N to mean a diseased allele and n to mean a normal allele, 5, 7 and 8 must be Nn because they have an unaffected parent. 2 is also Nn, as if it was NN all its progeny would be Nn and so affected by the disease, which is not the case, as 3 and 4 are unaffected.

Question 292: A

Since 6 is disease free, his genotype must be nn. Thus, neither of 6's parents could be NN, as otherwise 6 would have at least one diseased allele.

Question 293: A

Urine passes from the kidney into the ureter and is then stored in the bladder. It is finally released through the urethra.

Question 294: F

Deoxygenated blood from the body flows through the inferior vena cava to the right atrium where it flows to the right ventricle to be pumped via the pulmonary artery to the lungs where it is oxygenated. It then returns to the heart via the pulmonary vein into the left atrium into the left ventricle where it is pumped to the body via the aorta.

Question 295: E

During inspiration, the pressure in the lungs decreases as the diaphragm contracts, increasing the volume of the lungs. The intercostal muscles contract in inspiration, lifting the rib cage.

Question 296: D

Whilst A, B, C and E are true of the DNA code, they do not represent the property described, which is that more than one combination of codons can encode the same amino acid, e.g. Serine is coded by the sequences: TCT, TCC, TCA, TCG.

Question 297: B
The degenerate nature of the code can help to reduce the deleterious effects of point mutations. The several 3-nucleotide combinations that code for each amino acid are usually similar such that a point mutation, i.e. a substitution of one nucleotide for another, can still result in the same amino acid as the one coded for by the original sequence.

The degenerate nature of the code does little to protect against deletions/insertions/duplications, which will cause the bases to be read in incorrect triplets, i.e. result in a frame shift.

Question 298: D
The hypothalamus is the site of central thermoreceptors. A decrease in environmental temperature decreases sweat secretion and causes cutaneous vasoconstriction to minimise heat loss from the blood.

Question 299: A
The movement of carbon dioxide in the lungs and neurotransmitters in a synapse are both examples of diffusion. Glucose reabsorption is an active process, as it requires work to be done against a concentration gradient.

Question 300: F
Some enzymes contain other molecules besides protein, e.g. metal ions. Enzymes can increase rates of reaction that may result in heat gain/loss, depending on if the reaction is exothermic or endothermic. They are prone to variations in pH and are highly specific to their individual substrate.

Question 301: C
Segment area $= \frac{60}{360}\pi r^2 = \frac{1}{6}\pi r^2$

$\frac{x}{\sin 30°} = \frac{2r}{\sin 60°}$

$x = \frac{2r}{\sqrt{3}}$

Total triangle area $= 2 \times \frac{1}{2} \times \frac{2r}{\sqrt{3}} \times 2r = \frac{4r^2}{\sqrt{3}}$

Proportion covered: $\frac{1}{6}\pi r^2 / \frac{4r^2}{\sqrt{3}} = \frac{\sqrt{3}\pi}{24} \approx 23\%$

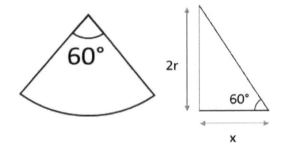

Question 302: B
$(2r)^2 = r^2 + x^2$

$3r^2 = x^2$

$x = \sqrt{3}r$

$Total\ height = 2r + x = (2 + \sqrt{3})r$

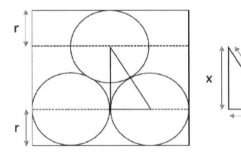

Question 303: A

$V = \frac{1}{3}h \times$ base area

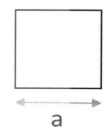
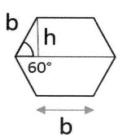

Therefore base area must be equal if h and V are the same

Internal angle = 180° – external ; external = 360°/6 = 60° giving internal angle 120°

Hexagon is two trapezia of height h where: $\frac{b}{\sin 90°} = \frac{h}{\sin 60°}$

$h = \frac{\sqrt{3}}{2}b$

Trapezium area $= \frac{(2b+b)}{2}\frac{\sqrt{3}}{2}b = \frac{3\sqrt{3}}{4}b^2$

Total hexagon area $= \frac{3\sqrt{3}}{2}b^2$

So from equal volumes: $a^2 = \frac{3\sqrt{3}}{2}b^2$

Ratio: $\sqrt{\frac{3\sqrt{3}}{2}}$

Question 304: C

A cube has 6 sides so the area of 9 cm cube = 6 x 9^2
9 cm cube splits into 3 cm cubes.
Area of 3 cm cubes = 3^3 x 6 x 3^2
$\frac{6 \times 3^2 \times 3^3}{6 \times 3^2 \times 3^2} = 3$

Question 305: E

$x^2 = (4r)^2 + r^2$

$x = \sqrt{17}r$

$\frac{\sqrt{17}r}{\sin 90°} = \frac{r}{\sin \theta}$

$\theta = \sin^{-1}\left(\frac{1}{\sqrt{17}}\right)$

Question 306: C

0 to 200 is 180 degrees so: $\frac{\theta}{180} = \frac{70}{200}$

$\theta = \frac{7 \times 180}{20} = 63°$

Question 307: C

Since the rhombi are similar, the ratio of angles = 1
Length scales with square root of area so length B = $\sqrt{10}$ length A

$\frac{angle\ A/angle\ B}{length\ A/length\ B} = \frac{1}{\sqrt{10}/1} = \frac{1}{\sqrt{10}}$

Question 308: E

$y = \ln(2x^2)$

$e^y = 2x^2$

$x = \sqrt{\frac{e^y}{2}}$

As the input is -x, the inverse function must be $f(x) = -\sqrt{\frac{e^y}{2}}$

Question 309: C

$log_8(x)$ and $log_{10}(x) < 0$; $x^2 < 1$; $\sin(x) \le 1$ and $1 < e^x < 2.72$

So e^x is largest over this range

Question 310: C

$x \propto \sqrt{z}^3$

$\sqrt{2}^3 = 2\sqrt{2}$

Question 311: A

The area of the shaded part, that is the difference between the area of the larger and smaller circles, is three times the area of the smaller so: $\pi r^2 - \pi x^2 = 3\pi x^2$. From this, we can see that the area of the larger circle, radius x, must be 4x the smaller one so: $4\pi r^2 = \pi x^2$

$4r^2 = x^2$

$x = 2r$

The gap is $x - r = 2r - r = r$

Question 312: D

$x^2 + 3x - 4 \ge 0$

$(x - 1)(x + 4) \ge 0$

Hence, $x - 1 \ge 0$ or $x + 4 \ge 0$

So $x \ge 1$ or $x \ge -4$

Question 313: C

$\frac{4}{3}\pi r^3 = \pi r^2$

$\frac{4}{3}r = 1$

$r = \frac{3}{4}$

Question 314: B

When $x^2 = \frac{1}{x}$; $x = 1$

When $x > 1, x^2 > 1, \frac{1}{x} < 1$

When $x < 1, x^2 < 1, \frac{1}{x} > 1$

Range for $\frac{1}{x}$ is $x > 0$

Non-inclusive so: $0 < x < 1$

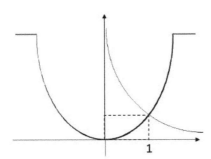

Question 315: A

Don't be afraid of how difficult this initially looks. If you follow the pattern, you get (e-e) which = 0. Anything multiplied by 0 gives zero.

Question 316: C

For two vectors to be perpendicular their scalar product must be equal to 0.

Hence, $\begin{pmatrix} -1 \\ 6 \end{pmatrix} \cdot \begin{pmatrix} 2 \\ k \end{pmatrix} = 0$

$\therefore -2 + 6k = 0$

$k = \frac{1}{3}$

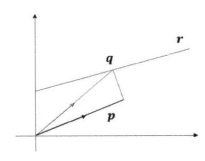

Question 317: C

The point, q, in the plane meets the perpendicular line from the plane to the point p.

$q = -3i + j + \lambda_1(i + 2j)$

$\overrightarrow{PQ} = -3i + j + \lambda_1(i + 2j) + 4i + 5j$

$= \begin{pmatrix} -7 + \lambda_1 \\ -4 + 2\lambda_1 \end{pmatrix}$

PQ is perpendicular to the plane r therefore the dot product of $\overrightarrow{PQ}$ and a vector within the plane must be 0.

$\begin{pmatrix} -7 + \lambda_1 \\ -4 + 2\lambda_1 \end{pmatrix} \cdot \begin{pmatrix} 1 \\ 2 \end{pmatrix} = 0$

$\therefore -7 + \lambda_1 - 8 + 4 + \lambda_1 = 0$

$\lambda_1 = 3$

$\overrightarrow{PQ} = \begin{pmatrix} -4 \\ 2 \end{pmatrix}$

The perpendicular distance from the plane to point p is therefore the modulus of the vector joining the two $\overrightarrow{PQ}$:

$|\overrightarrow{PQ}| = \sqrt{(-4)^2 + 2^2} = \sqrt{20} = 2\sqrt{5}$

Question 318: E

$-1 + 3\mu = -7 ; \mu = -2$

$2 + 4\lambda + 2\mu = 2 \therefore \lambda = 1$

$3 + \lambda + \mu = k \therefore k = 2$

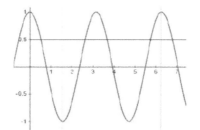

Question 319: E

$\sin\left(\frac{\pi}{2} - 2\theta\right) = \cos(2\theta)$

Root solution to $\cos(\theta) = 0.5$

$\theta = \frac{\pi}{3}$

Solution to $\cos(2\theta) = 0.5$

$\theta = \frac{\pi}{6}$

Largest solution within range is: $2\pi - \frac{\pi}{6} = \frac{(12-1)\pi}{6} = \frac{11\pi}{6}$

Question 320: A

$\cos^4(x) - \sin^4(x) \equiv \{\cos^2(x) - \sin^2(x)\}\{\cos^2(x) + \sin^2(x)\}$

From difference of two squares, then using Pythagorean identity $\cos^2(x) + \sin^2(x) = 1$

$\cos^4(x) - \sin^4(x) \equiv \cos^2(x) - \sin^2(x)$

But double angle formula says: $\cos(A + B) = \cos(A)\cos(B) - \sin(A)\sin(B)$

$\therefore$ if $A = B, \cos(2A) = \cos(A)\cos(A) - \sin(A)\sin(A)$

$= \cos^2(A) - \sin^2(A)$

So, $\cos^4(x) - \sin^4(x) \equiv \cos(2x)$

Question 321: C

Factorise: $(x + 1)(x + 2)(2x - 1)(x^2 + 2) = 0$

Three real roots at $x = -1, x = -2, x = 0.5$ and two imaginary roots at 2i and -2i

Question 322: C
An arithmetic sequence has constant difference d so the sum increases by d more each time:
$$u_n = u_1 + (n-1)d$$
$$\sum_1^n u_n = \frac{n}{2}\{2u_1 + (n-1)d\}$$
$$\sum_1^8 u_n = \frac{8}{2}\{4 + (8-1)3\} = 100$$

Question 323: E
$$\binom{n}{k} 2^{n-k}(-x)^k = \binom{5}{2} 2^{5-2}(-x)^2$$
$$= 10 \times 2^3 x^2 = 80x^2$$

Question 324: A
Having already thrown a 6 is irrelevant. A fair die has equal probability $P = \frac{1}{6}$ for every throw.

For three throws: $P(6 \cap 6 \cap 6) = \left(\frac{1}{6}\right)^3 = \frac{1}{216}$

Question 325: D

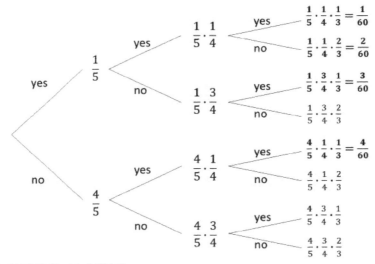

Total probability is sum of all probabilities:
$$= P(Y \cap Y \cap Y) + P(Y \cap Y \cap N) + P(Y \cap N \cap Y) + P(N \cap Y \cap Y)$$
$$= \frac{1}{60} + \frac{2}{60} + \frac{3}{60} + \frac{4}{60} = \frac{10}{60} = \frac{1}{6}$$

Question 326: C
$$P[(A \cup B)'] = 1 - P[(A \cup B)]$$
$$= 1 - \{P(A) + P(B) - P(A \cap B)\}$$
$$= 1 - \frac{2+6-1}{8} = \frac{3}{8}$$

Question 327: D
Using the product rule: $\frac{dy}{dx} = x \cdot 4(x+3)^3 + 1 \cdot (x+3)^4$
$$= 4x(x+3)^3 + (x+3)(x+3)^3$$
$$= (5x+3)(x+3)^3$$

Question 328: A

$\int_1^2 \frac{2}{x^2} dx = \int_1^2 2x^{-2} dx =$

$\left[\frac{2x^{-1}}{-1}\right]_1^2 = \left[\frac{-2}{x}\right]_1^2$

$= \frac{-2}{2} - \frac{-2}{1} = -1$

Question 329: D

Express $\frac{5i}{1+2i}$ in the form $a + bi$

$\frac{5i}{1+2i} \cdot \frac{1-2i}{1-2i}$

$= \frac{5i+10}{1+4} - \frac{5i+10}{5}$

$= i + 2$

Question 330: B

$7\log_a(2) - 3\log_a(12) + 5\log_a(3)$

$7\log_a(2) = \log_a(2^7) = \log_a(128)$

$3\log_a(12) = \log_a(1728)$

$5\log_a(3) = \log_a(243)$

This gives: $\log_a(128) - \log_a(1728) + \log_a(243)$

$= \log_a\left(\frac{128 \times 243}{1728}\right) = \log_a(18)$

Question 331: E

Functions of the form quadratic over quadratic have a horizontal asymptote.

Divide each term by the highest order in the polynomial i.e. x^2:

$$\frac{2x^2 - x + 3}{x^2 + x - 2} = \frac{2 - \frac{1}{x} + \frac{3}{x^2}}{1 + \frac{1}{x} - \frac{2}{x^2}}$$

$$\lim_{x \to \infty}\left(\frac{2 - \frac{1}{x} + \frac{3}{x^2}}{1 + \frac{1}{x} - \frac{2}{x^2}}\right) = \frac{2}{1} \quad i.e. \, y \to 2$$

So, the asymptote is $y = 2$

Question 332: A

$1 - 3e^{-x} = e^x - 3$

$4 = e^x + 3e^{-x} = \frac{(e^x)^2}{e^x} + \frac{3}{e^x} = \frac{(e^x)^2 + 3}{e^x}$

This is a quadratic equation in (e^x): $(e^x)^2 - 4(e^x) + 3 = 0$

$(e^x - 3)(e^x - 1) = 0$

So $e^x = 3, x = \ln(3)$ or $e^x = 1, x = 0$

Question 333: D

Rearrange into the format: $(x + a)^2 + (y + b)^2 = r^2$

$(x - 3)^2 + (y + 4)^2 - 25 = 12$

$(x - 3)^2 + (y + 4)^2 = 47$

$\therefore r = \sqrt{47}$

Question 334: C

$\sin(-x) = -\sin(x)$

$\int_0^a 2\sin(-x)\,dx = -2\int_0^a \sin(x)\,dx = -2[\cos(x)]_0^a = \cos(a) - 1$

Solve $\cos(a) - 1 = 0$ $\therefore a = 2k\pi$

Or simply the integral of any whole period of $\sin(x) = 0$ i.e. $a = 2k\pi$

Question 335: E

$\frac{2x+3}{(x-2)(x-3)^2} = \frac{A}{(x-2)} + \frac{B}{(x-3)} + \frac{C}{(x-3)^2}$

$2x + 3 = A(x - 3)^2 + B(x - 2)(x - 3) + C(x - 2)$

When $x = 3, (x - 3) = 0$, $C = 9$

When $x = 2, (x - 2) = 0, A = 7$

$2x + 3 = 7(x - 3)^2 + B(x - 2)(x - 3) + 9(x - 2)$

For completeness: Equating coefficients of x^2 on either side: $0 = 7 + B$ which gives: $B = -7$

Question 336: E

Forces on the ball are $weight = mg$ which is constant and tension T which varies with position.

$F = ma$; $a = \frac{v^2}{r}$

$T + mg = m\frac{v^2}{r}$

If the ball stops moving in a circle it means there is no tension in the string (T=0) so: $mg = m\frac{v^2}{r}$

$v = \sqrt{gr}$

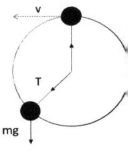

Question 337: E

To move at a steady velocity there is no acceleration, so the forces are balanced. Resolve along the slope: $F + mg\sin(30) = T\cos(30)$

$\frac{mg}{2} + F = \frac{T\sqrt{3}}{2}$

$T = \frac{2}{\sqrt{3}}\left(\frac{mg}{2} + F\right)$

Work done in pulling the box is W=Fd where the distance can be expressed as a function of velocity and time $d = v\Delta t$ so: $W = vF\Delta t$

Since power can expressed as $P = \frac{W}{\Delta t}$ we have $P = \frac{vF\Delta t}{\Delta t}$

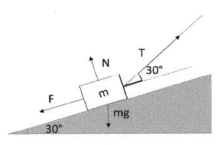

$P = vT\cos(30)$

$P = \left(\frac{mg}{2} + F\right)v$

Question 338: C

This is a conservation of energy problem. In the absence of friction there is no dissipation of energy therefore the sum of the potential and kinetic energy must be constant: $\frac{1}{2}mv^2 + mgh = E$

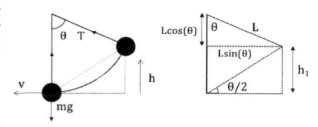

At its highest the velocity and kinetic energy are 0 so $E = mgh_1$.

At the bottom of the swing the potential energy at h is converted to kinetic energy.

Therefore: $\frac{1}{2}mv^2 = E = mgh_1 \therefore v = \sqrt{2gh_1}$

$h_1 = l(1 - \cos(\theta))$

$v = \sqrt{2gh_1} = \sqrt{2gl(1 - \cos(\theta))}$

Question 339: E

Conservation of momentum. Before : $p = mu_1$

Afterward: $p = m(v_1 + v_2)$

Vertical components must cancel since spheres are moving in opposite directions therefore: $v_2 \sin(\theta) = 2v_2 \sin(30°)$

$\sin(\theta) = 1$ giving $\theta = 90°$

Question 340: D

This is a modified Newton's pendulum. Elastic collision means that both kinetic energy and momentum are conserved. Therefore three balls will swing at a velocity equal to the velocity of the first to conserve both momentum and kinetic energy.

$$\frac{1}{2}mv^2 = \frac{1}{2}m_2u^2$$

But momentum is also conserved so $mv = m_2u$

Mass must therefore be equal i.e. 3 balls move at a velocity $u = v$

Question 341: A

The ball will follow a parabolic trajectory. The minimum angle is therefore given by the gradient of the parabola which goes through the points (0,3) and (-6,0). [**NB:** do not use a triangle, the ball lands 6m behind the fence so it takes a parabolic trajectory with the vertex directly above the fence. The smallest possible initial angle places the ball just passing over the fence.]

$y = -ax^2 + 3$

$0 = -36a + 3$

$a = \frac{1}{12}$

$y = -\frac{x^2}{12} + 3$

$\frac{dy}{dx} = -\frac{2x}{12} = -\frac{x}{6}$

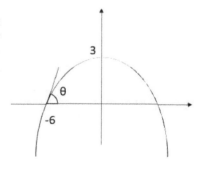

Question 342: B

Simple harmonic motion $m\frac{d^2x}{dt^2} = -kx$

Hence: $T \propto \sqrt{\frac{m}{k}}$.

Doubling k and halving m would therefore reduce time period T by a half. The frequency is the reciprocal of the time period and will therefore double.

Question 343: C

At the top of the bounce the kinetic energy is zero as velocity is zero. Highest velocity will be downwards before impact where *potential energy lost = kinetic energy gained* (assuming no air resistance and therefore conservation of energy).

$\frac{1}{2}mv^2 = mgh$

Kinetic energy before hitting the ground:

E_k=m.10.3 = 30m

Highest velocity:

$v^2 = 2gh = 60$

$v = 2\sqrt{15}$

Question 344: D

Speed is close to c so need to consider Lorentz contraction in special relativity: $l' = l\sqrt{1 - \frac{v^2}{c^2}}$

$= l\sqrt{1 - \frac{\left(\frac{c}{10}\right)^2}{c^2}}$

$= l\sqrt{0.99}$

Question 345: B

Initial kinetic energy must equal work done to stop the car: $\frac{1}{2}mv^2 = Fd = \frac{mg}{2}d$

$v^2 = gd$

$d = \frac{v^2}{g}$

Question 346: D

Find the proportion of amplitude left, then use this to work out how many half-lives have passed: $\frac{25}{200} = \frac{1}{8} = \frac{1}{2^3}$

Therefore 12 seconds is three half-lives and $t_{1/2}$=4s.

Question 347: C

$f_{beats} = |f_1 - f_2| = \frac{1}{8}f = 10$

$f = 80$ Hz

Question 348: E

The two waves would interfere destructively as they are half a wavelength phase difference. A wave would reflect back onto itself in this way if reflected in a plane, perpendicular surface. These two waves travelling in opposite directions (incident and reflected) would produce a standing wave, with this exact point in time corresponding to zero amplitude. There are 5 nodes with two fixed ends making it the 4th harmonic of a standing wave. Thus, all the statements are true.

Question 349: C

Beta decay changes a neutron to a proton releasing an electron and an antineutrino (*a* doesn't change, *b* increases by one), then alpha decay emits an alpha particle which is two protons and two neutrons (*a* decreases by 4, *b* decreases by two).

Question 350: A

Assume ideal gas: $PV = nRT$

$$P_2 = \frac{nR2T}{1.1V}$$

Therefore change in P is equal to $\frac{2}{1.1} = 1.818$ which is an 82% increase.

Question 351: A

Alpha particles are +2 and are deflected to the right. Were they a -1 charge they would follow path Q however electrons have a far smaller mass and will be deflected much more than an alpha particle. Then the beam of electrons is more likely to follow path P.

Question 352: D

If brightness increases with current, each association can be analysed individually and remembering that all the bulbs are identical.

For the association of three bulbs in the parallel: since the three bulbs are in parallel with the source of energy, the current leaving the source will be equally divided between each one of them and equal to $I = \frac{V}{R}$ For the association of three bulbs in series the same value of current passes throw three bulbs and is equal to the voltage divided by the total resistance $I = \frac{V}{3R}$

Finally, for the association of one bulb in parallel with two bulbs in series the current will be divided to the total resistance of each branch.

For two in series and one in parallel, the two in series have resistance 2R so: $\frac{1}{R_T} = \frac{1}{2R} + \frac{1}{R} \therefore R_T = \frac{2R}{3}$

$$I_T = \frac{V}{R_T} = \frac{3V}{2R}$$

This is split with a third going to the branch with resistance 2R and two thirds going to the branch with resistance R: $I_6 = \frac{V}{R}$; $I_{4,5} = \frac{V}{2R}$

Current in 1,2,3 and 6 is the same and is the greatest

Question 353: A

Object is between f and the lens, so rays will diverge on the other side producing a virtual image on the same side which is magnified.

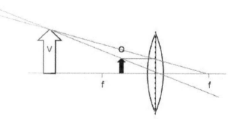

Question 354: B

Moments taken with the pivot at the wall must balance therefore:

$$\frac{2}{3}lT\sin\theta = lmg$$

$$T = \frac{3mg}{2\sin\theta}$$

Question 355: C

This process is known as the photoelectric effect (photons producing electron emission) and the presence of a work function arises due to wave particle duality. [n.b. thermionic emission uses heat not incident radiation to emit particles]. As the axis is kinetic energy and not potential, the intercept is the work function not the stopping potential.

Question 356: D

Huygens' principle states that every point on a wavefront is like a point source of a wave travelling at the same speed. This explains the first four but does not account for energy loss during propagation i.e. damping.

Question 357: A

Carnot cycle is the most efficient where: $\eta = \frac{work\ done}{heat\ put\ in} = 1 - \frac{T_{cold}}{T_{hot}} = 1 - \frac{240}{420} = \frac{3}{7} \approx 43\%$

Question 358: B

B is the only correct statement. NB- while generators can have a moving coil, they could equally have a moving magnetic field instead, so this is not true.

Question 359: C

NAND (gives X), OR (gives Y) and AND (gives Z) gates

Question 360: E

Several almost synonymous terms. Those in bold are correct. E is the only line with all three.

	P	Q	R
A	**Elastic Modulus**	**Yield stress**	Fracture toughness
B	**Tensile Modulus**	**Plastic onset**	Yield stress
C	Hardness	**Stiffness**	**Ductile failure**
D	Ductility	**Elastic limit**	Brittle fracture
E	**Young's Modulus**	**Yield stress**	**Fracture stress**

END OF SECTION

Section 2: Worked Answers

Physics Question 1

a)

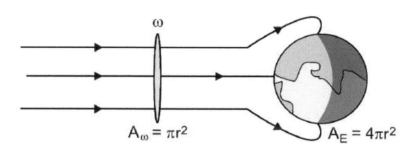

$$\omega_E = (A_\omega / A_E)\omega \quad = \frac{\omega}{4}$$

b) Radiation reflected by the atmosphere $= a\Omega/4$

Radiation emitted as infrared from the Earth's surface $= \sigma T_s^4$, where T_s is the temperature of the Earth's surface

c) Total solar radiation that reaches the Earth = radiation reflected by the atmosphere + radiation emitted as infrared from the Earth's surface, i.e.

$$\frac{\Omega}{4} = \frac{a\Omega}{4} + \sigma T_s^4$$

Rearranging for T_s: $\Omega = a\Omega + 4\,\sigma T_s^4$

$$T_s^4 = \frac{\Omega(1-a)}{4\sigma}$$

$$T_s = \sqrt[4]{\frac{\Omega(1-a)}{4\sigma}}$$

d) When $\Omega = 1372$ Wm^{-2} and a = 0.3, **T$_s$ = 255 K**

e) The black-body temperature of the Earth is far lower than the actual average surface temperature of the Earth (which is around 290 K). This is because the insulating effect of the atmosphere has not been considered in the calculations above. Effects of volcanism and other sources of heath within the Earth are also ignored in these calculations.

f) $d = 1.5 \times 10^{11}$ m, and $r_{sun} = 6.96 \times 10^8$ m and at d, $\Omega_E = 1372$ Wm^{-2}

The total power emitted per unit area from the sun = (ratio of the area of the sphere of radius d and r_{sun}) $\times \Omega_E$,

i.e. $\Omega_{Sun} = \left(\frac{4\pi d^2}{4\pi r_{sun}^2}\right)\Omega_E$

Therefore: $\Omega_{Sun} = \sigma T_{Sun}^4$

$$T_{Sun} = \sqrt[4]{\frac{\Omega_{Sun}}{\sigma}}$$

$$T_{Sun} = \sqrt[4]{\left(\frac{4\pi d^2}{4\pi r_{sun}^2}\right)\frac{\Omega_E}{\sigma}}$$

Using the values for d, r_{sun} & Ω_E, **T_{Sun} = 5790 K**

g) In this calculation, we have assumed a perfectly circular orbit and a constant solar power output.

Physics Question 2

a)

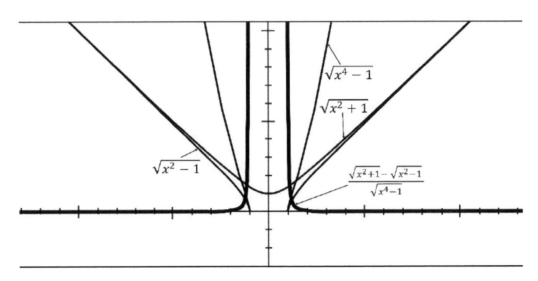

$\sqrt{x^4-1}$

$\sqrt{x^2+1}$

$\sqrt{x^2-1}$

$\dfrac{\sqrt{x^2+1}-\sqrt{x^2-1}}{\sqrt{x^4-1}}$

b) Evaluate: $\int \dfrac{\sqrt{x^2+1}-\sqrt{x^2-1}}{\sqrt{x^4-1}}\, dx$

Step 1 – Simplify the terms

The denominator is simplified as: $\sqrt{x^4-1} = (x^2+1)^{1/2}(x^2-1)^{1/2}$

Substituting this back into the original equation:

$$\int \frac{\sqrt{x^2+1}-\sqrt{x^2-1}}{\sqrt{x^4-1}}\, dx = \int \frac{\sqrt{x^2+1}-\sqrt{x^2-1}}{\sqrt{x^2+1}\sqrt{x^2-1}}\, dx$$

$$= \underbrace{\int \frac{1}{\sqrt{x^2-1}}\, dx}_{(I_1)} \; + \; \underbrace{\int \frac{1}{\sqrt{x^2+1}}\, dx}_{(I_2)}$$

Step 2 – Integration by substitution

I_1 and I_2 need to be integrated separately using different substitutions.

I_1 – Use the substitution $x = \tan(t)$

$$dx = \sec^2(t)dt = \frac{1}{\cos^2(t)}\, dt$$

Substituting $x = \tan(t)$ into $\sqrt{x^2+1}$ gives:

$$\sqrt{\tan^2(t)+1} = \sqrt{\sec^2(t)} = \sec(t) = \frac{1}{\cos(t)}$$

Putting this back into the original integral for I_1:

$$\int \frac{1}{\sqrt{x^2+1}}\, dx = \int \frac{1}{\sec(t)} \frac{1}{\cos^2(t)}\, dt = \int \cos(t) \frac{1}{\cos^2(s)}\, ds$$

$$I_1 = \int \frac{1}{\cos(t)}\, dt$$

I_2 – Use the substitution $x = \sec(s) = \dfrac{1}{\cos(s)}$:

$$dx = \sec(s)\tan(s)ds = \frac{1}{\cos(s)} \frac{\sin(s)}{\cos(s)}\, ds = \frac{\sin(s)}{\cos^2(s)}\, ds$$

Substituting $x = \sec(s)$ into $\sqrt{x^2 - 1}$ gives:

$$\sqrt{sec^2(s) - 1} = \sqrt{tan^2(s)} = tan(s) = \frac{sin(s)}{cos(s)}$$

Putting this back into the original integral for I₂:

$$\int \frac{1}{\sqrt{x^2-1}} dx = \int \frac{1}{\tan(s)} \frac{\sin(s)}{\cos^2(s)} ds = \int \frac{\cos(s)}{\sin(s)} \frac{\sin(s)}{\cos^2(s)} ds$$

$$I_2 = \int \frac{1}{\cos(s)} ds$$

Hence the total integral is : $I_1 + I_2 = \int \frac{1}{\cos(t)} dt - \int \frac{1}{\cos(s)} ds$

Using the Rule: $\int \frac{1}{\cos(A)} dA = ln|secA + tanA| + C$

$$I_1 + I_2 = \int \frac{1}{\cos(t)} dt - \int \frac{1}{\cos(s)} ds = ln|\sec(t) + \tan(t)| - ln|\sec(s) + \tan(s)| + C$$

Using the trig identity: $tan^2(A) + 1 = sec^2(A)$

We can express t and s in terms of x:
$x = tan(t)$
$tan^2(t) + 1 = sec^2(t)$
$x^2 + 1 = sec^2(t)$
$sec(t) + tan(t) = \sqrt{x^2 + 1} + x$

$x = \sec(s)$
$sec^2(s) - 1 = tan^2(s)$
$x^2 - 1 = tan^2(s)$
$sec(s) + tan(s) = \sqrt{x^2 - 1} + x$

Hence:

$$\int \frac{\sqrt{x^2 + 1} - \sqrt{x^2 - 1}}{\sqrt{x^4 - 1}} dx = ln\left|\sqrt{x^2 + 1} + x\right| - ln\left|\sqrt{x^2 - 1} + x\right| + C$$

c)

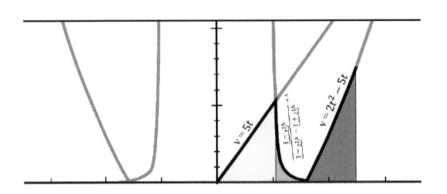

d)

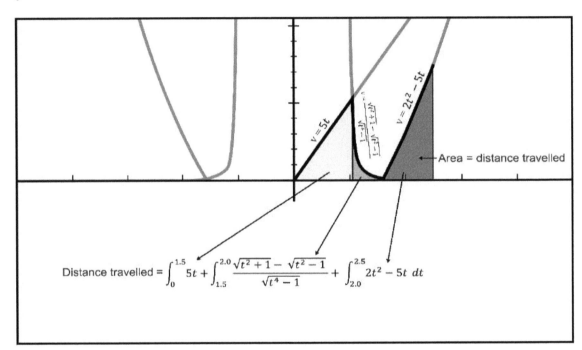

Distance travelled = $\int_0^{1.5} 5t + \int_{1.5}^{2.0} \frac{\sqrt{t^2+1} - \sqrt{t^2-1}}{\sqrt{t^4-1}} + \int_{2.0}^{2.5} 2t^2 - 5t \ dt$

Distance = $\int_0^{1.5} 5t + \int_{1.5}^{2.0} \frac{\sqrt{t^2+1} - \sqrt{t^2-1}}{\sqrt{t^4-1}} + \int_{2.0}^{2.5} 2t^2 - 5t \ dt$

Distance = $\left[\frac{5}{2}t^2\right]_0^{1.5} + \left[ln\left|\sqrt{x^2+1}+x\right| - ln\left|\sqrt{x^2-1}+x\right|\right]_{1.5}^{2.0} + \left[\frac{2}{3}t^3 - \frac{5}{2}t^2\right]_{2.0}^{2.5}$

Distance = $5.625 + 0.106 + 10.703 = \mathbf{16.44 \ km}$

Physics Question 3

a) At ∂N, $-\lambda \partial t$ atoms have decayed, $\frac{dN}{dt} = -\lambda N$ therefore $\int \frac{1}{N} dN = \int -\lambda dt$, $\ln(N) = -\lambda t + C$

When $t = 0$, $\ln(N_0) = C$ therefore $\ln(N) = -\lambda t + \ln(N_0)$ and $N = N_0 e^{-\lambda t}$

b) $A = A_0 e^{-\lambda t}$

c) At $t = t_{1/2}$, $\frac{N_0}{2} = N_0 e^{-\lambda t_{1/2}}$ therefore $N_0 = 2N_0 e^{-\lambda t_{1/2}}$ and $1 = 2e^{-\lambda t_{1/2}}$

$-\lambda t_{1/2} = \ln\left(\frac{1}{2}\right)$

$t_{1/2} = \frac{1}{\lambda}\ln(2)$

i.e. $t_{1/2} = \frac{0.693}{\lambda}$

di)

Parent Isotope	Daughter Isotope	Decay Products	λ (a^{-1})
^{238}U	^{206}Pb	$8 \propto + 6\beta^-$	1.55×10^{-10}
^{235}U	^{207}Pb	$7 \propto + 4\beta^-$	9.85×10^{-10}
^{232}Th	^{208}Pb	$6 \propto + 4\beta^-$	4.95×10^{-11}
^{87}Rb	^{87}Sr	β^-	1.42×10^{-11}
^{147}Sm	^{143}Nd	$\propto$	6.54×10^{-12}
^{40}K	^{40}Ca and ^{40}Ar	β^- and electron capture	4.95×10^{-10} and 5.81×10^{-11}
^{39}Ar	^{39}Ar	β^-	2.57×10^{-3}
^{176}Lu	^{176}Hf	β^-	1.94×10^{-11}
^{187}Re	^{187}Os	β^-	1.52×10^{-11}
^{14}C	^{15}N	β^-	1.21×10^{-4}

dii) If the total number of daughter atoms present at time t is D, then the total number of atoms remaining, $N = N_0 - D$, and therefore $D = Ne^{-\lambda t}$

$N_0 - D = N_0 e^{-\lambda t}$

Becomes $D = N_0(1 - e^{-\lambda t})$

Using $N = N_0 e^{-\lambda t}$

$D = (e^{-\lambda t} - 1)$

Therefore $e^{-\lambda t} = \frac{D}{N} + 1$

$-\lambda t = \ln\left(\frac{D}{N} + 1\right)$

$t = -\frac{1}{\lambda}\ln\left(\frac{D}{N} + 1\right)$

diii) Age of the Earth: ^{238}U (or any isotope with a half-life on the order of the age of the Earth)

Ancient artefacts: ^{14}C, due to its short half-life

e) Assuming that $[^{206}\text{Pb}] \gg [^{208}\text{Pb}]$, we can assume that all Pb present in the sample is due to the decay of ^{238}U

The amount of ^{238}U present now is

$[^{238}U_T] = \frac{11.7 \times 10^{-5}}{238} = 4.92 \times 10^{-7}$ moles of ^{238}U

When one mole of ^{238}U decays, one mole of ^{206}Pb is produced, therefore the total ^{206}Pb present = amount of ^{238}U that has decayed

$$\left[^{236}Pb_T\right] = \frac{3.58 \times 10^{-5}}{206} = 1.74 \times 10^{-7} \text{ moles of } ^{206}\text{Pb}$$

The total ^{238}U that was present at time $t = 0$ must be $\left[^{238}U_0\right] = \left[^{238}U_T\right] + \left[^{236}Pb_T\right] = 6.66 \times 10^{-7}$ moles

As $t_{1/2} = \frac{0.693}{\lambda_{238}}$ and $\lambda_{238} = \frac{0.693}{4.5 \times 10^9} = 1.54 \times 10^{-10}$

Hence as $N = N_0 e^{-\lambda t}$,

$t = -\frac{1}{\lambda} ln\left(\frac{N}{N_0}\right)$

Where $N = \left[^{238}U_T\right]$ therefore $t = -\frac{1}{1.54 \times 10^{-10}} ln\left(\frac{4.92 \times 10^{-7}}{6.66 \times 10^{-7}}\right)$

$\underline{\boldsymbol{t = 1.97 \times 10^9 \text{ years}}}$

Physics Question 4

ai) $V = -\frac{Gm_1 m_2}{r}$

aii) $a = -\frac{\partial V}{\partial r} = -\frac{\partial}{\partial r}\left(\frac{Gm_1}{r}\right) = -\frac{Gm_1}{r^2}$

bi)

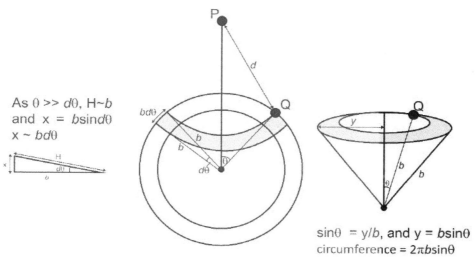

As $\theta \gg d\theta$, H~b
and $x = b\sin d\theta$
$x \sim bd\theta$

$\sin\theta = y/b$, and $y = b\sin\theta$
circumference $= 2\pi b\sin\theta$

Area of strip = circumference x width
Area $= (2\pi b\sin\theta)(bd\theta)$

$\boxed{\text{Area} = 2\pi b^2 \sin\theta d\theta}$

bii) Mass of strip = area of strip x thickness x density

$M = 2\pi b^2 \sin\theta d\theta t\rho$

At point Q, as $V = -\frac{Gm}{r}$

$V = -\frac{2G\pi b^2 \sin\theta d\theta t\rho}{D}$

Distance D can be calculated as

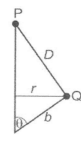

Using the cosine rule that
$a^2 = b^2 + c^2 - 2bc\cos A$,
$D^2 = b^2 + r^2 - 2br\cos\theta$

As $V = -G\int_m \frac{dm}{r}$ for a distribution of masses,

$V = -G\int_m = -\frac{2\pi b^2 \sin\theta d\theta t\rho}{D}$

$V = -2G\pi b^2 t\rho \int_m = -\frac{\sin\theta}{\sqrt{b^2+r^2-2rb\cos\theta}}d\theta$

However, $D^2 = b^2 + r^2 - 2brCos\theta$

$2DdD = 2brSin\theta d\theta$

$sin\theta d\theta = \frac{D}{br}dD$

Substituting this back into the original equation gives:

$V = -2G\pi b^2 t\rho \int_m = -\frac{D}{Dbr}dD$

$V = -2G\pi b^2 t\rho \int_m = -\frac{1}{br}dD$

$V = -2G\pi b^2 t\rho \left[\frac{D}{br}\right]_{D_{min}}^{D_{max}}$

There are two cases for point P, when P is inside and outside the hollow shell. When P is inside the shell, D_{max} and D_{min} are $b + r$ and $b - r$ respectively, therefore:

$$V_{inside} = -2G\pi b^2 t\rho \left[\frac{D}{br}\right]_{b-r}^{b+r} = -4G\pi b t\rho$$

When P is outside the hollow shell, D_{max} and D_{min} are $r + b$ and $r - b$ respectively, therefore:

$$V_{outside} = -2G\pi b^2 t\rho \left[\frac{D}{br}\right]_{r-b}^{r+b} = \frac{-4G\pi b^2 t\rho}{r}$$

biii) When P is inside the shell, the potential V is constant and independent of position

biv) As the potential V inside the shell is constant, acceleration a is zero

bv) $a = -\frac{\partial V_{outside}}{\partial r} = -\frac{\partial}{\partial r}\left(\frac{4G\pi b^2 t\rho}{r}\right)$

$a = -\frac{G(4\pi b^2 t\rho)}{r^2}$

$a = -\frac{GM}{r^2}$

Physics Question 5

ai) The production of electromotive forces (emfs) and currents caused by a changing magnetic field through a metal coil is called electromagnetic induction

aii) In the upper case, as the N pole of a magnet approaches face A of the coil, face A becomes a North pole by inducing an anticlockwise current in the coil in order to oppose the forward motion of the magnet's N pole into the coil.

In the lower case, the N pole of a magnet moves away from face A of a coil. A clockwise current is induced in the coil, making face A South pole in order to oppose the motion of the magnet's N pole out of the coil.

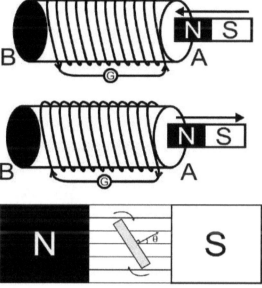

aiii) Faraday's law states that when magnetic flux changes through a circuit, an emf is induced for as long as the change in magnetic flux continues. This causes an emf, F, to be produced when a metal coil spins inside a magnetic field as shown below

$$F = BACos\theta$$

For a coil with N turns, $FN = BANCos\theta$

Where B = magnetic field strength and A = area of coil. The magnetic flux is given by $\emptyset = BA$, and Lenz's law states $V = -N\frac{\Delta\emptyset}{\Delta t}$

Where V = voltage of induced current, and t = time. When motion of the coil ceases, $\frac{\Delta\emptyset}{\Delta t} = 0$ and the induced current becomes zero, therefore satisfying Faraday's law.

aiv) Area of the coil = 50 cm², and as $V = -N\frac{\Delta\emptyset}{\Delta t}$

$$V_{gen} = -35\frac{0.4\times(0.05\times0.1)}{50/60}$$

$$\boldsymbol{V_{gen} = -8.4\ mV}$$

bi) The dynamo theory describes the process through which a rotating, convecting and electrically conducting fluid acts to maintain a magnetic field. In the Earth, the conductive fluid is the Earth's metallic outer core, and is made of primarily Fe and Ni. Heat produced from the radioactive decay of isotopes in the core is the source of energy that keeps the temperature of the outer core above the melting temperature of Fe, Ni, etc.

bii) Incoming charged particles from the solar wind would not be deflected by the Earth's magnetic field and would ionise on the Earth's surface. As a result, an atmosphere would not be stable, nor would the presence of liquid water on the Earth's surface.

The absence of a magnetic field indicates the absence of a rotating core of liquid metal (i.e. a metallic planetary outer core) required to generate a current. This will eventually occur when the activity of radioactive nuclides in the Earth's core is too low to sustain the temperatures required for a stable liquid outer core, causing the outer core to freeze and solidify. As a result, dynamo motion will cease, and the planetary magnetic field will be lost.

biii) Io, a Jovian moon. Volcanism is caused by heat produced by tidal heating due to the gravitational attraction of Jupiter and is not caused by an internally decaying metallic core as is the case in the Earth.

Chemistry Question 1

a) When at chemical equilibrium, the rates of both the forward and backward reactions are equal. The concentrations of the reactants and products may not necessarily be equal but they will be constant when the reaction is at equilibrium, i.e. equilibrium may lie on one side of the reaction.

b) $CO_{2(g)} \rightleftharpoons CO_{2(aq)}$ (1)

$H^+_{(aq)} + CO_3^{2-}{}_{(aq)} \rightleftharpoons HCO_3^-{}_{(aq)}$ (2)

$HCO_3^-{}_{(aq)} + H^+_{(aq)} \rightleftharpoons H_2CO_{3(aq)}$ (3)

The full reaction for the dissolution of CO_2 into seawater is actually:

$$CaCO_{3\,(s)} + CO_{2(aq)} + H_2O \rightleftharpoons Ca^{2+}_{(aq)} + 2HCO_3^-_{(aq)}$$

Where $CaCO_{3(s)}$ is limestone/carbonates in the ocean

c) From the equation, $\frac{[H_2CO_3]}{p[CO_2]} = 10^{-1.47}$, we can write $[H_2CO_3] = 10^{-1.47}\, p[CO_2]$
Therefore $[H_2CO_3] = 10^{-1.47}\,(340 \times 10^{-6}) = \mathbf{10^{-4.94}}$

d)
$$[H^+] = [HCO_3^-] + 2[CO_3^{2-}] + [OH^-]$$

Using the answer for part c, we know that $[H_2CO_3] = 10^{-4.94}$, therefore

$$\frac{[H^+][HCO_3^-]}{[H_2CO_3]} = 10^{-6.35}$$

becomes

$$[H^+][HCO_3^-] = 10^{-6.35}[H_2CO_3] = (10^{-6.35})(10^{-4.94}) = 10^{-11.29}$$

Hence

$$[HCO_3^-] = \frac{10^{-11.29}}{[H^+]} \quad (1)$$

Using the equations in the table and the relationship given in part b, we can write:

And

$$\frac{[H^+][CO_3^{2-}]}{[HCO_3^-]} = 10^{-10.33} \quad (2)$$

Substituting the value of $[HCO_3^-]$ from (1) into (2) we get:

$$[H^+][CO_3^{2-}] = 10^{-10.33}[HCO_3^{2-}]$$
$$[CO_3^{2-}] = 10^{-10.33}\frac{10^{-11.29}}{[H^+]}$$
$$[CO_3^{2-}] = \frac{10^{-21.62}}{[H^+]^2} \quad (3)$$

From the table we can also write that:

$$[OH^-] = \frac{10^{-14}}{[H^+]} \quad (4)$$

Substituting (1), (3) and (4) into the original equation gives us

$$[H^+] = [HCO_3^-] + 2[CO_3^{2-}] + [OH^-]$$

$$[H^+] = \frac{10^{-11.29}}{[H^+]} + \frac{10^{-21.32}}{[H^+]^2} + \frac{10^{-14}}{[H^+]}$$

e) Assuming that as $10^{-11.29} \ll 10^{-14} \ll 10^{-21.62}$, the answer for part d becomes

$$[H^+] \approx \frac{10^{-11.29}}{[H^+]}$$

Therefore

$$[H^+]^2 \approx 10^{-11.29}$$
$$[H^+] \approx 10^{-5.65}$$

Or

$$[H^+] \approx 2.2\,\mathbf{ppm}$$

f) $pH = -\log([H^+])$
g) $pH = -\log(10^{-5.65}) = \mathbf{5.65}$

h) Firstly, as

$$\frac{[H_2SO_3]}{p[SO_2]} = 10^{-0.096} \text{ and } [SO_2] = 0.2\text{ppb(v)}, [H_2SO_3] = 10^{-0.096}p[SO_2] = 10^{-0.096}(0.2 \times 10^{-9})$$

$$[H_2SO_3] = 10^{-9.603}$$

Using the same method and answer to part d,

$$[H^+] = [HCO_3^-] + 2[CO_3^{2-}] + [HSO_3^-] + 2[SO_3^{2-}] + [OH^-]$$

becomes

$$[H^+] = \frac{10^{-11.29}}{[H^+]} + \frac{10^{-21.32}}{[H^+]^2} + [HSO_3^-] + 2[SO_3^{2-}] + \frac{10^{-14}}{[H^+]}$$

Where $[HSO_3^-]$ and $2[SO_3^{2-}]$ are unknown. From the equations and values given in the table, we can write:

$$\frac{[H^+][HSO_3^-]}{[H_2SO_3]} = 10^{-1.77}$$

Therefore

$$[H^+][HSO_3^-] = 10^{-1.77}[H_2SO_3]$$

Substituting in the value of $[H_2SO_3]$ from above,

$$[H^+][HSO_3^-] = (10^{-1.77})(10^{-9.603})$$
$$[H^+][HSO_3^-] = 10^{-11.37}$$
$$[HSO_3^-] = \frac{10^{-11.37}}{[H^+]} \quad (1)$$

And

$$\frac{[H^+][HSO_3^{2-}]}{[SO_3^{2-}]} = 10^{-7.21}$$

Therefore

$$[H^+][SO_3^-] = 10^{-7.21}[HSO_3^-]$$
$$[H^+][SO_3^-] = 10^{-7.21}\frac{10^{-11.37}}{[H^+]}$$

And

$$[SO_3^-] = 10^{-7.21}\frac{10^{-11.37}}{[H^+]^2}$$

$$[SO_3^-] = \frac{10^{-18.58}}{[H^+]^2} \quad (2)$$

Finally, substituting (1) and (2) into the equation

$$[H^+] = \frac{10^{-11.29}}{[H^+]} + \frac{10^{-21.32}}{[H^+]^2} + [HSO_3^-] + 2[SO_3^{2-}] + \frac{10^{-14}}{[H^+]}$$

Gives:

$$[H^+] = \frac{10^{-11.29}}{[H^+]} + \frac{10^{-21.32}}{[H^+]^2} + \frac{10^{-11.37}}{[H^+]} + \frac{10^{-18.26}}{[H^+]^2} + \frac{10^{-14}}{[H^+]}$$

As $10^{-11.29} + 10^{-11.37} \gg 10^{-14} \gg 10^{-18.26} \gg 10^{-21.32}$,

$$[H^+] \approx \frac{10^{-11.29}}{[H^+]} + \frac{10^{-11.37}}{[H^+]}$$

Hence

$$[H^+]^2 \approx 10^{-11.29} + 10^{-11.37}$$
$$[H^+] \approx 10^{-5.51}$$

Therefore

$$\underline{\boldsymbol{pH \approx 5.51}}$$

i) The main sources of atmospheric SO_2 are volcanoes and anthropogenic sources

Chemistry Question 2

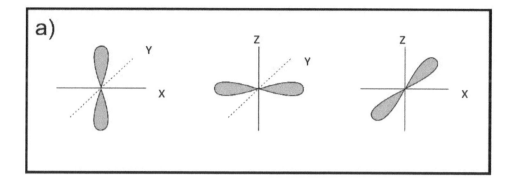

bi)

$$XH_3 + H^- \rightarrow XH_4^- \quad i.e.$$

$$XH_3 + H^+ \rightarrow XH_4^+ \quad i.e.$$

$$XH_3^- + H^+ \rightarrow XH_4 \quad i.e.$$

bii)

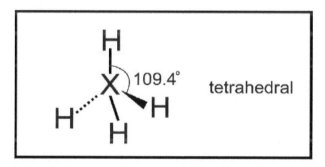

biii) X can be C, N, or B

ci)

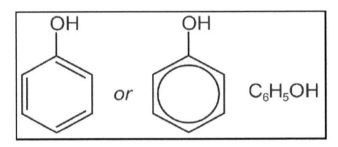

cii) Reaction occurring is $C_6H_{12}O_2 \rightleftarrows C_6H_{11}O_2^- + H^+$,

therefore, $\frac{[H^+][C_6H_{11}O_2^-]}{[C_6H_{12}O_2]} = 10^{-pk_a}$

taking -log on both sides: $-\log_{10}\left[\frac{[H^+][C_6H_{11}O_2^-]}{[C_6H_{12}O_2]}\right] = pk_a$ (1)

Looking at the reaction equation $[C_6H_{11}O_2^-] = [H^+] = 10^{-pH} = 10^{-3.2} = 6.3 \times 10^{-4} M$

As all $C_6H_{11}O_2^-$ is made from $C_6H_{12}O_2$: $[C_6H_{12}O_2] + [C_6H_{11}O_2^-] = 0.05$ M

$$[C_6H_{12}O_2] = 0.05 \text{ M} - [C_6H_{11}O_2^-] = 4.94 \times 10^{-2} \text{ M}$$

Plugging in values into the equation (1):

$$pk_a = -\log_{10}\left[\frac{(6.3 \times 10^{-4}) \times (6.3 \times 10^{-4})}{(4.94 \times 10^{-2})}\right] = 5.09$$

ciii)

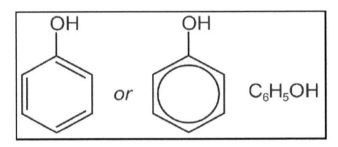

To simplify this problem, we state that pKa of 2,2 di-methyl butanoic acid is significantly lower than pKa of phenol (5.09 vs. 7.5), meaning that 2,2 di-methyl butanoic acid is relatively stronger acid. We will make an assumption that 2,2 di-methyl butanoic acid will be closer to its equilibrium because it is a relatively stronger acid.

Therefore, we first calculate 2,2 di-methyl butanoic acid equilibrium [H+] using ICE table:
$$C_6H_{12}O_2 \rightleftarrows C_6H_{11}O_2^- + H^+$$

I	0.1M	0	0
C	-x	+x	+x
E	0.1-x	x	x

$$10^{-pKa} = Ka = \frac{[H^+][C_6H_{11}O_2^-]}{[C_6H_{12}O_2]}$$

Therefore, $10^{-5.09} = 8.13 \cdot 10^{-6} = \frac{x \cdot x}{0.1-x}$

And after solving a quadratic equation: $[H^+] = x = 2.89 \cdot 10^{-4}$ M

Now we use the calculated $[H^+]$ value to calculate phenol equilibrium using ICE table:

$$C_6H_5OH \rightleftarrows C_6H_5O^- + H^+$$

I	0.1M	0	$2.89 \cdot 10^{-4}$ M
C	-x	+x	+x
E	0.1-x	x	$2.89 \cdot 10^{-4}$ M + x

$$10^{-pKa} = Ka = \frac{[H^+][C_6H_5O^-]}{[C_6H_5OH]}$$

Therefore, $10^{-7.5} = 3.16 \cdot 10^{-8} = \frac{x \cdot (2.89 \cdot 10^{-4} \text{ M} + x)}{0.1-x}$

Solving the quadratic equation gives us:

$$x = 4.36 \cdot 10^{-5} \text{ M}$$

As in the ICE table: $[H^+] = 2.89 \cdot 10^{-4}$ M $+ x = 2.89 \cdot 10^{-4}$ M $+ 4.36 \cdot 10^{-5}$ M $= 3.323 \cdot 10^{-4}$ M

Therefore $pH = -\log[H^+] \approx 3.48$

Note: remember that we got this pH value after an assumption that 2,2 di-methyl butanoic acid will dominate in an equilibrium, therefore the value we get is an approximation, and hence the approximation sign used.

Chemistry Question 3

a)

(i) 3-dimethyl butane

(ii) Heptanal

(iii) Methyl-propanoate

(iv) Ethanenitrile

(iv) 2-bromo,3-chloro butane

bi) (Thermal) Cracking

bii) $H_3C - CH_2 - CH_2 - CH_2 - CH_2 - CH_2 - CH_2 - CH_2 - CH_2 - CH_3 \rightarrow H_3C - CH = CH_2 + H_3C - CH_2 - CH_2 - CH_2 - CH_2 - CH_2 - CH_2$
i.e. decane $\rightarrow$ propane + heptane (one possible answer)

biii) Gasoline, diesel oil, bitumen, lubricants, petroleum gases, kerosene

biv) Zeolites

bv) Necessary conditions: UV light

1. Initiation
   $Cl_2 \rightarrow 2\ Cl\cdot$
2. Propagation
   $Cl\cdot + C_2H_6 \rightarrow HCl + C_2H_5\cdot$
   $Cl_2 + C_2H_5\cdot \rightarrow C_2H_5Cl + Cl\cdot$
3. Termination
   $Cl\cdot + Cl\cdot \rightarrow Cl_2$ <u>or</u>
   $Cl\cdot + C_2H_5\cdot \rightarrow C_2H_5Cl$ <u>or</u>
   $C_2H_5\cdot + C_2H_5\cdot \rightarrow C_4H_{10}$

bvi) The most common impurity found in hydrocarbons is sulphur. The greatest risk during the thermal cracking process is combustion/explosion

Chemistry Question 4

ai) Isomerism is the phenomenon whereby certain compounds with the same molecular formula, exist in different forms due to different organisations of atoms.

aii)

There are 6 isomers of C_4H_8

aiii) Molecules (i), (ii), (iii), (iv), (v) and (vi) are all structural isomers as they are all the same chemical formula however atoms are arranged differently. Molecules (ii) and (iii) are stereoisomers, as atoms are in the same sequence of chemical bonds, yet in a different arrangement, i.e. (ii) = cis and (iii) = trans.

bi) and **bii)**

(i) Ethene + Hydrogen Bromide

iii) Cyclohexene + Hydrogen Bromide

biii) Electrophilic addition

Chemistry Question 5

a) The bond strength of H-X compounds varies and depends on X. The reactivity of H-X molecules is inversely proportional to the bond strength, such that molecule reactivity decreases from H-I, H-Br, H-Cl to H-F, where H-I has the weakest bond (therefore most reactive) and H-F has the strongest bond (therefore least reactive).

b)

(i)

Intermediary

$$CH_3CH_2-O^- \quad K^+$$

Final products

$$+ \; H-Br \; (or \; Br^- \; or \; KBr) \; + \; H_3C-CH_2-OH$$

(ii)

Intermediary 1

Intermediary 2

$$CH_3CH_2-O^-$$

Final products

$$+ \; H-Br \; (or \; Br^- \; or \; KBr) \; + \; H_3C-CH_2-OH$$

ci)

$$k_a = \frac{[CH_3COO^-][H^+]}{[CH_3COOH]}$$

$$pk_a = -log_{10}[k_a]$$

cii) Because there are only two components, $[CH_3COO^-]\sim[H^+]$ and therefore

$$k_a = \frac{[CH_3COO^-][H^+]}{[CH_3COOH]} \sim \frac{[H^+]^2}{[CH_3COOH]}$$

Therefore

$[H^+] = \sqrt{10^{-pk}[CH_3COOH]} = \sqrt{10^{-4.76}\times 0.017} = 5.435\times 10^{-4}$

and $pH = -log_{10}(5.435\times 10^{-4}) = \mathbf{3.26}$

ciii)

Ethanol

$H_3C-CH_2-OH \rightleftharpoons H_3C-CH_2-O^- + H^+$

Ethanoic acid

Both reactions involve ionising an O-H bond, however for CH_3COO^- the O^- anion is stabilised by resonance/delocalisation

Equilibrium therefore lies on the RHS of this reaction. This is not the case for O^- anions in alcohols, this making carboxylic acids more acidic than alcohols

Biology Question 1

a) DNA has to be translated to mRNA, pairing the 3' DNA strand with a 5' mRNA strand. This leads to the following:

DNA: 3' GAC ACG CCG AGT 5'
mRNA: 5' CUG UGC GGC UCA 3'
AA: Leu Cys Gly   Ser

b) Yes, a mutation of this base could retain the same amino acids as according to the decoding wheel, the UUG base triplet also encodes Leucine.

c) The gene sequence could be cloned from the genome and inserted into a plasmid using restriction ezyme digestion and ligation. The plasmid would need to use appropriate bacterial promoters in order to ensure gene transcription. Finally, the plasmid can be inserted into bacteria using various methods, such as heat shock, electroporation or chemical transformation, allowing the expression and production of insulin.

d) . There are multiple possible explanations. The most likely is that the mutation occurred in the amplification protein, causing too much insulin to be released. However, it is possible that the receptor had mutated to have an increased affinity for glucose, the receptor had mutated to be able to signal constitutively to the 'amplification' protein, and feedback systems to control insulin release had mutated to release more insulin. The more amplification occurs, the larger the amount of insulin that is released.

e) Type 2 diabetes is marked by cells experiencing reduced 'effective' insulin concentrations. This can represent one scenario. Firstly, due to sustained high levels of insulin in the blood, the cells become desensitized to insulin, effectively experiencing reduced insulin effect. As the cells begin requiring an increasing amount of insulin to achieve the same degree of stimulation and Glut-4 translocation, the pancreas has to continuously increase insulin production until, at some point, the pancreas can no longer keep up with insulin production and this will result in a decrease in effective insulin concentrations, decreased translocation of Glut-4 transporter to the cell surface which in turn will reduce the amount of glucose that enters the cell.

f) Glucagon opposes the action of insulin. With regards to the liver, this means that glucagon increases the mobilisation of glucose from hepatic storages. As glucose is stored in form of glycogen, glucagon will cause an increase in glycogenolysis which will allow glucose to be released raising the blood sugar levels.

Biology Question 2

a) Vertebrates have myelinated nerves, which increases the transmission speed of information.

b)

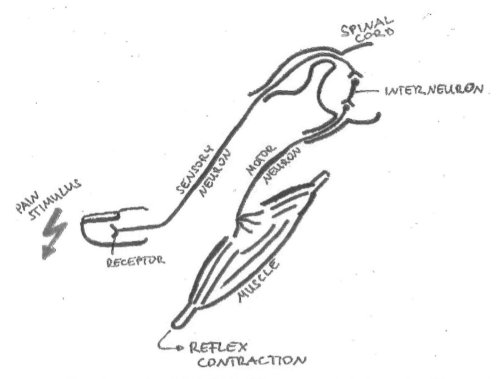

c) The purpose of reflexes is protection of the individual. It serves to cue the body to act quickly to withdraw the affected body part from danger. The purpose of transmission of the information to the brain lies in the higher computing ability of the brain thereby allowing for more precise analysis of the stimulus.

d) Mimicry is the idea that non-toxic species copy the external attributes of toxic species as defence mechanisms. This serves to protect them from predators. Usually mimicry involves adoption of bright colours in distinctive patterns, as many toxic species have bright colours in distinctive patterns.

e) Action potentials are generated by the coordinated action of voltage-gated ion channels in response to neurotransmitter or other stimulus. This initial stimulus depolarises the membrane to its threshold potential, at which point sodium ions (Na+) flow in through voltage-gated sodium channels, depolarising the membrane and propagating the action potential along the neurone. In response, voltage-gated potassium channels open and the sodium channels close, leading to efflux of potassium ions (K+). This causes repolarisation and then hyperpolarisation of the membrane, preventing it from firing again for a time (known as the refractory period). The Na+/K+ transporter can then act to return the cell to its resting membrane potential.

f) Lidocaine blocks sodium channels, thereby blocking the progress of the action potential along the nerve fibre as the sodium channels are needed to generate the electric current to produce depolarisation of the nerve membrane.

Biology Question 3

a) In large units, leadership becomes essential. There is a strict separation between alpha animals and other members of the group. With the alpha status comes the primary right for breeding as well as other advantages with regards to food acquisition etc. The hierarchical order is maintained by constant competition meaning that only the strongest individuals remain in power.

b) Organisation of individuals into groups leads to a variety of advantages including protection from predators as well as separation of labour as exemplified here in the success rates of hunting depending on hunting party size. There is also the evolutionary advantage resulting from the constant competition for the leadership positions which ensures that only the strongest individuals proliferate leading to propagation of beneficial genes through the population.

c) Communication allows for the control of large groups of individuals as it allows transmission of orders as well as communication of emotions, etc. Generally speaking, one has to differentiate between different types of communication. There is chemical communication in insects and then there are forms of verbal communication in higher apes. The complexity of communication processes requires an increasing degree of cerebral complexity.

d) The reproductive pressure is important as it again ensures that the strongest individuals in the herd have the greatest proliferative success. In addition to that, proliferation is always connected to external environments with breeding periods being timed to ensure the maximal likelihood of offspring survival.

e) Genetic diversity helps maintain genetic adaptability and protects the population from accumulation of negative genetic traits such as genetic diseases. Maintenance of genetic adaptability is important as it allows populations to adjust to a variety of different environments.

f) Female dominance exists in several animal species such as lions, elephants and bonobo apes. However, male dominance is more common. This could be due to factors such as physical superiority and aggression. According to the so-called 'prior attribute hypothesis' dominance hierarchies are supposed to arise from intrinsic attributes which are pre-existing individual differences in strength, such as body mass, age, sex or physiological traits. There is much empirical evidence for the winner-loser effect: It appears to operate in many species, ranging from insects, crustaceans, fishes, amphibians, and reptiles, to mammals.

Biology Question 4

a) Ruminants regurgitate their food and therefore digestion of food matter is more complete meaning that toxins accumulate to a higher degree than it does in monogastriers.

b)

➢ Heart: Increase of rate and contractility, also increase of energy demand

➢ Lungs: bronchodilation, increase of respiratory rate, increase of energy demand

➢ Gut: reduction in peristalsis, reduction in perfusion, reduction in nutrient absorption

➢ Blood vessels: increase in blood pressure, centralisation of blood, contraction of peripheral blood vessels, dilation of blood vessels in skeletal muscle and heart

c) N-Methyltyramine acts by competing with adrenaline for breakdown by MAO-A. Increase in adrenergic stimulation causes a general increase in energy demand. Paired with the inhibition of citric acid cycle by fluoroacetate this leads to complete consumption of energy reserves leading to death.

d) Competitive antagonists act by occupying the active site of the enzyme, thus rendering it unable to carry out its usual function until the inhibitor is released. In this case, it would reduce the amount of MAO-A available to metabolise adrenalin at any given moment, contributing to higher adrenalin levels and consequential resource consumption.

e) Part of the solution would be to increase activity of MAO-A as this would maintain appropriate breakdown adrenaline. Moreover, fluoroacetate will inhibit the citric acid cycle. Administration of drugs designed to stimulate metabolism, in order to overcome the effects of fluroacetate, and drugs designed to reduce the production of adrenalin, to counteract N-methyltyramine would be effective treatments. Additionally, administration of glucose/sugars as an easily metabolised source of energy and possibly the induction of vomiting to reduce the amount of toxin absorbed are other alternative treatments.

f) Fight or flight acts to increase the body's readiness to physical performance either with regards to fighting or to running away to withdraw from the danger. This is aided by the selective increase in performance of all systems supporting skeletal muscle contraction (Increased heart rate, vasodilatation in skeletal muscle as it is important to allow increased blood flow and therefore nutrient delivery to and waste removal from the muscles, and bronchodilation) and the shut-down of all non-performance oriented process.

Biology Question 5

a)

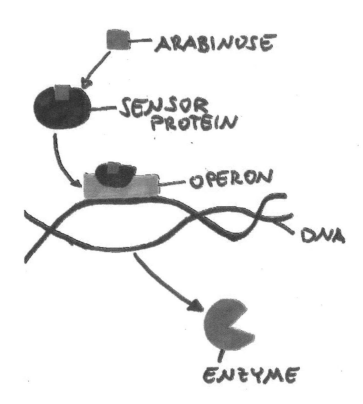

b) The response arc works along the same lines. A protein complex senses the presence of lactose in the environment and through that triggers the expression of lactase.

c) A gain of function mutation in the operon would lead to constantly activated expression of the associated digestive enzyme. This would result in the permanent production of this enzyme which is problematic as is will cause continuous energy consumption.

d) Bacteria are unable to control the general make up of their environment. For this reason, they they live a 'feast or famine'-type existence. Being adaptable to the nutrients provided by the environment ensures that no resources are wasted on unnecessary enzymes.

e) This means that certain bacterial species have acquired or developed sensing mechanisms that monitor the concentration of antibiotics in the environment leading to the expression of countermeasures in response to antibiotic presence. This is important as it shifts the sensitivity patterns of bacteria to antibiotics in sometimes unpredictable directions.

f) As humans, we can influence our environment due to the complex nature of our interaction with it. This means that we can take control over what resources we are provided with making facultative expression less necessary. One example of facultative expression in humans would be lactase expression in the gut. Deactivation of expression will lead to lactose intolerance.

Final Advice

Arrive well rested, well fed and well hydrated

The NSAA is an intensive test, so make sure you're ready for it. Ensure you get a good night's sleep before the exam (there is little point cramming) and don't miss breakfast. If you're taking water into the exam then make sure you've been to the toilet before so you don't have to leave during the exam. Make sure you're well rested and fed in order to be at your best!

Move on

If you're struggling, move on. Every question has equal weighting and there is no negative marking. In the time it takes to answer on hard question, you could gain three times the marks by answering the easier ones. Be smart to score points- especially in section 2 where some questions are far easier than others.

Afterword

Remember that the route to a high score is your approach and practice. Don't fall into the trap that *"you can't prepare for the NSAA"*– this could not be further from the truth. With knowledge of the test, some useful time-saving techniques and plenty of practice you can dramatically boost your score.

Work hard, never give up and do yourself justice.

Good luck!

Acknowledgements

I would like to express my sincerest thanks to the many people who helped make this book possible, especially the Oxbridge Tutors who shared their expertise in compiling the huge number of questions and answers.

Rohan

About UniAdmissions

UniAdmissions is an educational consultancy that specialises in supporting **applications to Medical School and to Oxbridge**.

Every year, we work with hundreds of applicants and schools across the UK. From free resources to our *Ultimate Guide Books* and from intensive courses to bespoke individual tuition – with a team of **300 Expert Tutors** and a proven track record, it's easy to see why UniAdmissions is the **UK's number one admissions company**.

To find out more about our support like **NSAA tuition** check out www.uniadmissions.co.uk/NSAA

Your Free Book

Thanks for purchasing this Ultimate Guide Book. Readers like you have the power to make or break a book – hopefully you found this one useful and informative. If you have time, *UniAdmissions* would love to hear about your experiences with this book.

As thanks for your time we'll send you another ebook from our Ultimate Guide series absolutely FREE!

How to Redeem Your Free Ebook in 3 Easy Steps

1) Find the book you have either on your Amazon purchase history or your email receipt to help find the book on Amazon.

2) On the product page at the Customer Reviews area, click on 'Write a customer review'

Write your review and post it! Copy the review page or take a screen shot of the review you have left.

3) Head over to www.uniadmissions.co.uk/free-book and select your chosen free ebook! You can choose from:
- ✓ The Ultimate NSAA Guide – 400 Practice Questions
- ✓ NSAA Practice Papers
- ✓ NSAA Past Paper Worked Solutions
- ✓ The Ultimate Oxbridge Interview Guide
- ✓ The Ultimate UCAS Personal Statement Guide

Your ebook will then be emailed to you – it's as simple as that!

Alternatively, you can buy all the above titles at **www.uniadmissions.co.uk/our-books**

NSAA Online Course

If you're looking to improve your NSAA score in a short space of time, our **NSAA Online course** is perfect for you. The NSAA Online Course offers all the content of a traditional course in a single easy-to-use online package – available instantly after checkout. The online videos are just like the classroom course, ready to watch and re-watch at home or on the go and all with our expert Oxbridge tuition and advice.

You'll get full access to all of our NSAA resources including:

- ✓ Copy of our acclaimed book "The Ultimate NSAA Guide"
- ✓ Full access to extensive NSAA online resources including:
- ✓ 2 Full Mock Papers
- ✓ Full worked solutions for all NSAA past papers
- ✓ 600 practice questions
- ✓ 6 hours online on-demand lecture series
- ✓ Ongoing Tutor Support until Test date – never be alone again.

The course is normally £99 but you can get **£ 20 off** by using the code *"UAONLINE20"* at checkout.

https://www.uniadmissions.co.uk/product/nsaa-online-course/

£20 VOUCHER:

UAONLINE20

9 781912 557066